Transforming Management with AI, Big-Data, and IoT

Fadi Al-Turjman • Satya Prakash Yadav
Manoj Kumar • Vibhash Yadav
Thompson Stephan
Editors

Transforming Management with AI, Big-Data, and IoT

Editors
Fadi Al-Turjman
Department of Artificial Intelligence Engineering
AI and Robotics Institute, Near East University
Mersin, Turkey

Manoj Kumar
School of Computer Science
University of Petroleum and Energy Studies
Dehradun, Uttar Pradesh, India

Thompson Stephan
Department of Computer Science
and Engineering, Faculty of Engineering and Technology
M. S. Ramaiah University of Applied Sciences
Bangalore, Karnataka, India

Satya Prakash Yadav
Department of Computer Science
and Engineering
G. L. Bajaj Institute of Technology
and Management (GLBITM),
Affiliated to AKTU
Greater Noida, India

Vibhash Yadav
Department of Information Technology
Rajkiya Engineering College
Banda, Uttar Pradesh, India

ISBN 978-3-030-86748-5   ISBN 978-3-030-86749-2   (eBook)
https://doi.org/10.1007/978-3-030-86749-2

This Springer imprint is published by the registered company Springer Nature Switzerland AG
The registered company address is: Gewerbestrasse 11, 6330 Cham, Switzerland

Preface

This book discusses the effect that artificial intelligence (AI) and Internet of Things (IoT) have on industry. The authors start by showing how the application of these technologies has already stretched across domains such as law, political science, policy, and economics and how it will soon permeate areas of autonomous transportation, education, and space exploration, only to name a few. The authors then discuss applications in a variety of industries. Throughout the volume, the authors provide detailed, well-illustrated treatments of each topic with abundant examples and exercises. This book provides relevant theoretical frameworks and the latest empirical research findings in various applications. The book is written for professionals who want to improve their understanding of the strategic role of trust at different levels of the information and knowledge society, that is, trust at the level of the global economy, of networks and organizations, of teams and work groups, of information systems, and, finally, of individuals as actors in the networked environments. It presents research in various industries and how artificial intelligence and Internet of Things are changing the landscape of business and management. It presents new and innovative features in artificial intelligence and IoT use and in raising economic efficiency at both micro and macro levels. Moreover, it examines case studies with tried-and-tested approaches for resolution of typical problems in each application of study.

Mersin, Turkey   Fadi Al-Turjman
Greater Noida, India   Satya Prakash Yadav
Dehradun, Uttar Pradesh, India   Manoj Kumar
Banda, Uttar Pradesh, India   Vibhash Yadav
Bangalore, Karnataka, India   Thompson Stephan

Contents

Artificial Intelligence for Smart Data Storage in Cloud-Based IoT

Pushpa Singh, Narendra Singh, P. Rama Luxmi, and Ashish Saxena

1 Introduction

Artificial intelligence (AI), the Internet of Things (IoT), and cloud computing are buzzwords in the modern technological era. AI is the technology that aims at making computers or machines equivalent to the human brain and thus capable of learning and problem-solving [27]. AI-based applications can be integrated easily with other emerging technologies like IoT, cloud, Big Data, and Blockchain [12]. IoT states a system of interrelated, connected objects or things that can collect and transfer data via the Internet. A substantial number of physical things are being associated with the Internet at an exceptional rate recognizing the concept of the IoT. Reports and recent trends show that there are more than 30 billion IoT connections, almost four IoT devices per person on average by the year 2025 (https://iot-analytics.com/state-of-the-iot-2020-12-billion-iot-connections-surpassing-non-iot-for-the-first-time/). These IoT applications generate massive data, and cloud

P. Singh (✉)
Department of Computer Science and Information Technology, KIET Group of Institutions, Delhi-NCR, Ghaziabad, India
e-mail: pushpa.kiet@kiet.edu

N. Singh
Department of Management Studies, GL Bajaj Institute of Management and Research, Greater Noida, India
e-mail: narendra.singh@glbimr.org

P. R. Luxmi
School of Electrical Engineering, Vellore Institute of Technology, Chennai, India
e-mail: sriramalakshmi.p@vit.ac.in

A. Saxena
Department of Management Studies, Sharda University, Greater Noida, India
e-mail: ashish.saxena2@sharda.ac.in

© The Author(s), under exclusive license to Springer Nature Switzerland AG 2022
F. Al-Turjman et al. (eds.), *Transforming Management with AI, Big-Data, and IoT*, https://doi.org/10.1007/978-3-030-86749-2_1

1

computing delivers a way for those generated data to travel to their endpoint [5]. The adoption of cloud computing is recognized as a data-processing and storage facility. All real-time applications connected with IoT need just-in-time processing and quick action over the clouds. AI and IoT-based data have obtained much attention from researchers, academicians, and industrialists in health care, agriculture, telecommunication, e−/m-commerce, and transportations. Nowadays, AI-based approaches amplify the role of IoT in business monitoring, health-care monitoring, disease prediction, bioinformatics, research and development, stock market prediction, social network analysis, weather analysis, agriculture, transportation, and resource optimization. Implementation of these applications requires data storage and computational capacity generally provided by cloud-based services [13]. AI techniques are used to process the stored data in a high-precision and just-in-time manner. The cloud is a powerful tool for transmitting data through the traditional Internet channels as well as via a devoted direct link. IoT becomes the source of generating huge data, and the clouds become crucial for data storage [9]. Hence, the IoT and clouds are closely integrated to offer commercial business services and generally referred to as cloud-based IoT. Businesses like Amazon Web Services (AWS), Google, and Microsoft have become certain cloud-based IoT services leaders, making the challenge even more worthwhile. Further, cloud-based IoT is used to connect a wide range of smart things in various applications.

AI, IoT, and cloud computing play significant roles in various aspects in the present and in the future too. AI methods aim to gather data from various industries to process and collect the data generated from cloud-based IoT. Integration of AI, IoT, and cloud has transformed the overall storage capacity and digital world [24] and hence has become a hot topic for all researchers and academicians. This chapter aims to emphasize on the role of AI in cloud and IoT-based data storage.

The remainder of the chapter is systematized as follows: Section 2 focuses on cloud-based data storage. Section 3 discusses the role of IoT in clouds. Further, the role of AI in IoT and cloud data storage is introduced in Sect. 4. Section 5 explains the applications of AI, IoT, and clouds in various sectors, and Sect. 6 concludes the chapter.

2 Cloud-Based Data Storage

Cloud storage is an Internet-based storage system in which data are transmitted on remote storage systems. The data generated from IoT and other devices are stored, maintained, managed, backed up, and accessible to users via the Internet. Users usually pay according to their consumption of cloud storage on a monthly basis. The cloud-based primary services are database services, computing services, and storage services. There are four basic types of cloud storage: public, private, hybrid, and community cloud data storage, as shown in Fig. 1.

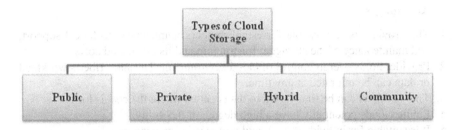

Fig. 1 Types of cloud storage

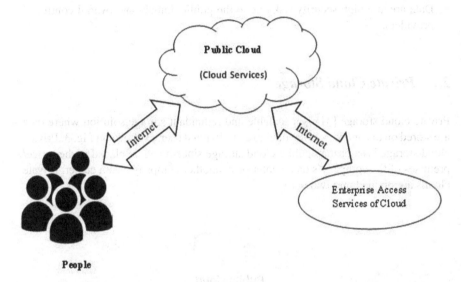

Fig. 2 Public cloud storage

2.1 Public Cloud Storage

Public cloud storage, also called Software as Services (SaaS), is provisioned for open use by the general public. Public cloud storage may be owned, managed, and functioned by a commercial, a government, or an academic. It offers data storage on a pay-per-use basis. In public cloud storage, the communication links can be assumed to be realized over the public Internet, as shown in Fig. 2. Any number of people, clients, organizations, or businesses can access cloud services by the mean of the public Internet. Examples of public cloud storage are Google App Engine, Microsoft Windows Azure, IBM Smart Cloud, Amazon Elastic Compute Cloud (EC2), etc.

Advantages

1. The provider is responsible for establishment, maintenance, technical support, and maintenance of the storage infrastructure and its associated costs.
2. Provider holds overall control of cloud environment. The subscriber's workload or data can be migrated at any time.
3. The workload can be transferred to data centers where the cost is low.
4. Public clouds potentially have a high degree of flexibility.
5. It is suitable for individual users and mid-sized organizations.

Limitation

- Data are at a high security risk due to the public domain and overall control of providers.

2.2   *Private Cloud Storage*

Private cloud storage [31] is a scalable and redundant storage solution where data are stored on distant servers devoted to an individual user, as shown in Fig. 3. Private cloud storage is safer than public cloud storage since it can be placed in the office/premises of the company's data center or in another company's data center. Private clouds are divided into two parts:

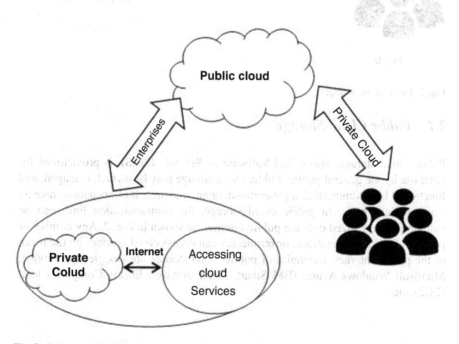

Fig. 3 Private cloud storage

On-Site Private Cloud—This type of private cloud is implemented at the user's locations.

Outsourced Private Cloud—This type of private cloud is employed at the server side, which is applied to a hosting business.

Advantages of Private Cloud Storage

1. The main benefit of private cloud storage is that it permits the user to have complete control over their data.
2. It offers high security, scalability, and reliability.
3. It is suitable for large enterprises or organizations.

Limitation
- It is expensive compared with public cloud storage.

2.3 Hybrid Cloud

Hybrid cloud combines private cloud (on-premises or off-premises) and public cloud as represented in Fig. 4. They have substantial deviations in performance, reliability, and security properties depending upon the type of cloud chosen to build

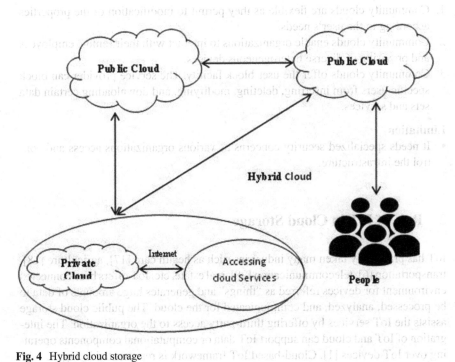

Fig. 4 Hybrid cloud storage

a hybrid cloud. Examples of hybrid clouds are Windows Azure, VMware, and vCloud.

Advantages

1. Customization is easy.
2. It is secure and reliable.
3. It is suitable for small and mid-size companies.

Limitation
• It consists of the limitation of both public and private clouds.

2.4 Community Cloud

Community cloud storage is the deviation of private cloud storage and exclusively useful for a particular community of users from organizations that have common concerns. Any data is stored in the community cloud owned as private cloud storage to manage the community's security. It offers a unique chance for businesses to work on common assignments. Google Apps for Government and Microsoft Government Community Cloud are some well-known community clouds.

Advantages

1. Community clouds are flexible as they permit to modification of the properties according to the user's needs.
2. Community clouds enable organizations to interact with their remote employees and provision of diverse heterogeneous devices.
3. Community clouds offer the user block facility. The service provider can block specific users from inserting, deleting, modifying, and downloading certain data sets and services.

Limitation
• It needs specialized security concerns as various organizations access and control the infrastructure.

3 Role of IoT in Cloud Storage

IoT has practically taken many industries such as health care [17], agriculture [18], transportation [6], telecommunication [28], real estate, etc. IoT offers best-connected environment for devices referred as "things" and generates huge amounts of data to be processed, analyzed, and communicated for the cloud. The public cloud storage assists the IoT services by offering third-party access to the organization. The integration of IoT and cloud can support IoT data or computational components operating over IoT devices [1]. Cloud-based IoT framework is represented in Fig. 5.

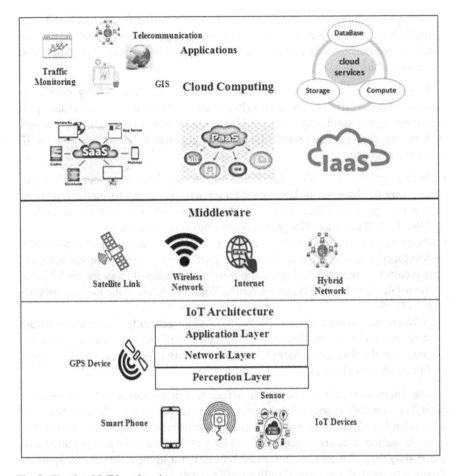

Fig. 5 Cloud and IoT-based environments

In Fig. 5, there are three components of cloud-based IoT storage. The first component provides IoT-based infrastructure where devices of different applications are connected with clouds using the second component, referred as middleware, which consists of communication technology such as 5G, Internet, Satellite Network, Wi-Fi, etc. The third component is the cloud infrastructure.

IoT-based Infrastructure: IoT-based infrastructure is based on a three-layered architecture, that is, perception layer, network layer, and application layer [7, 22].

- *The perception layer* is the physical layer or sensor layer of the architecture. The sensors and actuators are collecting data from things and transmit data for further processing [32]. It senses specific physical parameters or identifies other smart objects in the location.
- *The network layer* is liable for linking to objects (things), network devices, and servers.

- The *application layer* is the uppermost layer where IoT can be deployed. The application layer is accountable for conveying application-specific facilities to the users such as smart homes, smart transportation, smart agriculture, and smart health.

Cloud-based infrastructure offers cloud-based services such as infrastructure as a service (IaaS), platform as a service (PaaS), and software as a service (SaaS) [14] for extending the application smarter. Users usually pay according to their usage, such as central processing unit (CPU) per hour, gigabyte (GB) storage per hour, IP usage per hour, etc.

- *Infrastructure as a service (IaaS)*: IaaS offers virtualized computing infrastructure over the Internet. In IaaS cloud-based services, it can be paid for services like storage, networking, servers, and virtualization. The example of IaaS is AWS EC2, Rackspace, Google Compute Engine (GCE), etc.
- *Platform as a service (PaaS)*: PaaS is a model in which cloud vendors offer developers a platform for building apps. PaaS provides a run-time environment to create, test, run, and deploy the application. Examples of PaaS are AWS Elastic Beanstalk, Heroku, and Windows Azure. Windows Azure is the most commonly used PaaS.
- *Software as a service (SaaS)*: By SaaS, one can generally access the software from any device, at any time, without installation of any software on your computer via the Internet. Examples of SaaS are BigCommerce, Google Apps, Salesforce, Dropbox, etc.

The foremost benefit of placing the IoT system in a cloud is that it offers more flexibility, scalability, and reliability to connected applications. Sometimes businesses don't need consistent requirements of data storage; it may be occasional. For example, online sales are increased particularly at the time of Diwali or Christmas; hence, only for 1–2 weeks, organizations require extra storage and computing infrastructure to meet all businesses' online requirements.

4 Role of AI in Smart Data Storage

Due to digitalization and recent technological advancements, data are generated enormously. About 90% of the world's data are generated just during the last couple of years [15]. These generated data can produce interesting patterns, meaningful information, and correlation if stored and processed efficiently. Earlier, these data were stored in data centers. In a data center, data are mostly stored on the premises of the business organization. Some data centers may be in a distributed location and not accessible in an efficient manner. Here, the cloud-based solution comes into the picture. The cloud is entirely off-premises, and entire data are accessible from anywhere and at any place via the Internet. IoT and clouds are making data storage as "smart." The IoT produces vast amounts of data, and cloud computing conveys a pathway for those data to travel to their destination. Data storage requirements of big industries are also high. These industries mostly rely on hybrid clouds due to the potential advantages of hybrid clouds such as elasticity, agility, cost-efficiency, etc. [20].

Before cloud storage, AI task was very expensive due to large data, hardware, and software requirements. The potential of cloud computing makes AI capabilities in a highly accessible manner. AI-based technologies and algorithms focus on the data to discover patterns or models that can explain or inform and predict. AI offers a new business dimension with cloud computing (cloud data storage) and assists corporations/businesses to organize their data, discover interesting patterns and correlation, deliver customer experiences, and improve workflows. Just imagine about a driverless car where a high level of accuracy is driven by AI and required to process data in terabytes just for a single autonomous car, but there is no provision to store our terabyte to exabyte data before processing. Here, cloud comes into the picture. But, if it does not have the ability to calculate accurate precision on time, then it is impossible to get exact route information timely for the driverless autonomous car. Here, AI takes the leading role. Vehicle-to-vehicle information is also required to avoid accidents, and here, it is important to realize the role of IoT. Overall, a driverless car or self-driving car cannot work without coordinating all these three buzzwords, that is, AI, cloud, and IoT.

It is significant to deal and lead with data before applying AI techniques for businesses. The following are the benefits of AI with cloud computing.

1. AI tools are used to enhance data management. Nowadays, businesses generate and collect massive repositories of data, and AI tools can process data management.
2. AI tools can support and streamline the way data are ingested, updated, and managed to provide precise real-time data to users, predict fraudulent, and identify the risk areas.
3. The customer relationship management (CRM) platform Salesforce and its Einstein AI tool are used to manage and enhance the CRM.
4. Within a cloud environment, AI requires historical data to identify patterns and trends and makes better recommendations for the customer.
5. Leveraging AI with clouds is a cost-efficient initiative. The cloud permits organizations to purchase only the storage they actually need and when they need.

AI and cloud-based storage system are transforming the business at every level. AI with cloud services enhance the AI based application and practices into the business so that the service providers can respond quickly, according to the market's competitive environment in advance. IoT-based cloud assists in renovating business and changing the world with AI.

5 Applications of AI, IoT, and Cloud in Various Sectors

In the recent age, emerging technologies such as AI, IoT, and clouds are at the bleeding edge of cost optimization and improved quality. AI and IoT have already proven themselves in different areas and sectors like health care, agriculture, transportation, e-commerce, and telecommunication [29] as shown in Fig. 6. In order to work with real-time applications, cloud computing is extended as fog computing that can satisfy the need of time-critical applications [30]. The applications of the AI with IoT and clouds in various sectors are accelerated the implementation of Industry 4.0 [34]. These emerging technologies are ready to adapt to the industry changes in real time and maximize the industry turnover.

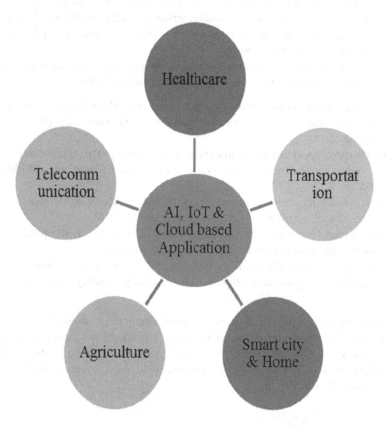

Fig. 6 Various AI, IoT, and cloud-based applications

5.1 Health-Care Sector

The health-care industry is generating enormous data that are stored in the cloud. Remote patients are connected via IoT sensor devices such as body temperature sensor, an electrocardiogram (ECG) sensor, blood pressure sensor, activity tracker, continuous glucose monitor (CGM), etc. [11]. These devices generate huge data from different locations. Hence, these devices need a cloud for continuous monitoring of specific attributes like blood glucose level, heartbeat, etc., for numerous days at a time by taking evaluations at regular intervals. AI and related techniques like machine learning and deep learning techniques are serving in timely analysis and prediction of diagnosis and prognosis of the disease [25]. Cloud assists in securing patient data and permitting health-care providers to continue conveying advanced technological care. A diversity of public, private, and hybrid cloud platforms supports getting better access to the patient's records and facilitating coordination between the doctor and patient.

5.2 Agriculture Sector

Automation of the agricultural system has enhanced the soil's gain and strengthened the fertility of the soil [3]. AI and IoT-based agriculture systems have various nodes such as soil, farmer, weather, irrigation, fertilizer, and crop management. These nodes or agents are like to distribute node that is connected through IoT, stored in a cloud environment for AI analytics. The various sensors are used for monitoring agricultural system. For examples, DHT11 is a low-cost digital temperature and humidity sensor, applied in the agricultural systems, and soil moisture sensor is applied for determining the soil moisture level. Cloud-based IoT is very supportive of integrating all agricultural-related data, such as soil-related data, weather data, crop-related data, farmers, supply chain, fertilizers, and retailers in the cloud [8]. The overall objective is to increase the productivity, prediction, and estimation of farming parameters to enhance economic efficiency.

5.3 Transportation Sector

Currently, transportation automation is the foremost area and emergent topic. IoT, AI, and machine learning techniques are profound for smart lighting systems and smart parking applications [35], as shown in Fig. 7. Moreover, route optimization, parking, and accident/detection seem to be the most popular applications. These applications produce and require a huge amount of data to be stored and processed. Cloud storage fulfills all these requirements of the smart transportation sector. Integration of AI, IoT, and clouds in transportation sectors makes this smart and also referred to as intelligent transport system (ITS) [16].

Transportation

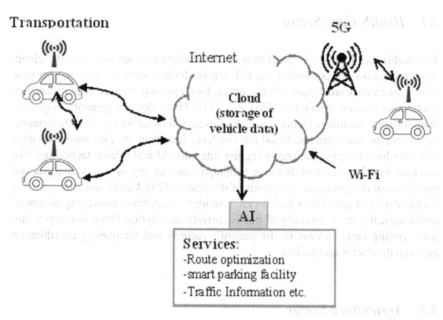

Fig. 7 Transportation with cloud, IoT, and AI

5.4 Telecommunication Sector

Telecommunication is itself facing a problem of poor quality of service (QoS), network selection criteria among multiple options and customer churning problems [4]. Telecommunication and mobile users want to always have the best-connected network. In fact, network infrastructure is a primary aspect of all emerging technologies. Telecommunication sector provides fastest network infrastructure to transmit the huge amount of data generated by AI based application. Emerging technologies also assist telecommunication infrastructure in searching best network, intelligent network selection, resource optimization, and user retention mechanism. Telecommunication network also leveraged with AI for intelligent network selection, resource optimization, and on-demand storage facility offered by cloud computing [23].

5.5 Smart City and Home

Smart city and smart home play a significant role in innovation trends [33]. AI and IoT are combinable with clouds and make the city and home smart [10]. Per and Skouby (2016) suggested an IoT architecture that incorporates smart homes and smart cities via the Cloud of Things (CoT) [19]. CoT virtualizes the IoT that offers monitoring and control [2]. AI, IoT, and clouds will contribute to the smart city/

home services developments. Concepts of the smart city, smart hospitals, smart education, smart governance, smart banking, and smart business [21] rely on AI, IoT, and cloud-based data storage. The collection vending machine (CVM) for e-waste management is associated with AWS cloud in order to make a city smart [26].

6 Conclusions

This chapter outlined the potential of AI and IoT in cloud-based data storage system. Recently, IoT-based applications are created by billions of connected devices. These devices are generating huge data, demanding huge data, and processing massive data. Various types of clouds such as public, private, hybrid, and community clouds fulfill the requirement of storage, processing, and maintenance of multiple applications. Further, the role of AI is also explained for analyzing and predicting the response to the IoT-based applications. In addition, cloud provides a place where one can implement the AI-based techniques for data analytics to make predictions about real-time behaviors of various applications such as health care, agriculture, transportation, telecommunication, and smart city/home.

References

1. Atlam, H. F., Alenezi, A., Alharthi, A., Walters, R. J., & Wills, G. B. (2017, June). Integration of cloud computing with internet of things: Challenges and open issues. In *2017 IEEE international conference on internet of things (iThings) and IEEE green computing and communications (GreenCom) and IEEE cyber, physical and social computing (CPSCom) and IEEE smart data (SmartData)* (pp. 670–675). IEEE.
2. Alohali, B., Merabti, M., & Kifayat, K. (2014, September). A cloud of things (cot) based security for home area network (han) in the smart grid. In *2014 eighth international conference on next generation mobile apps, services and technologies* (pp. 326–330). IEEE.
3. Al-Sakran, H. O. (2015). Framework architecture for improving healthcare information systems using agent technology. *International Journal of Managing Information Technology, 7*(1), 17.
4. Ahmad, A. K., Jafar, A., & Aljoumaa, K. (2019). Customer churn prediction in telecom using machine learning in big data platform. *Journal of Big Data, 6*(1), 28.
5. Bhardwaj, A., Al-Turjman, F., Kumar, M., Stephan, T., & Mostarda, L. (2020). Capturing-the-invisible (CTI): Behavior-based attacks recognition in IoT-oriented industrial control systems. *IEEE Access*, 1. https://doi.org/10.1109/ACCESS.2020.2998983
6. Bui, K. H. N., & Jung, J. J. (2019). ACO-based dynamic decision making for connected vehicles in IoT system. *IEEE Transactions on Industrial Informatics, 15*(10), 5648–5655.
7. Burhan, M., Rehman, R. A., Khan, B., & Kim, B. S. (2018). IoT elements, layered architectures and security issues: A comprehensive survey. *Sensors, 18*(9), 2796.
8. Choudhary, S. K., Jadoun, R. S., & Mandoriya, H. L. (2016). Role of cloud computing technology in agriculture fields. *Computing, 7*(3), 1–7.
9. Chithaluru, P., Al-Turjman, F., Kumar, M., & Stephan, T. (2020). I-AREOR: An energy-balanced clustering protocol for implementing green IoT in smart cities. *Sustainable Cities and Society*. https://doi.org/10.1016/j.scs.2020.102254

10. Dubey, S., Singh, P., Yadav, P., & Singh, K. K. (2020). Household waste management system using IoT and machine learning. *Procedia Computer Science, 167*, 1950–1959.
11. Dias, D., & Paulo Silva Cunha, J. (2018). Wearable health devices—Vital sign monitoring, systems and technologies. *Sensors, 18*(8), 2414.
12. Gill, S. S., Tuli, S., Xu, M., Singh, I., Singh, K. V., Lindsay, D., … Pervaiz, H. (2019). Transformative effects of IoT, Blockchain and artificial intelligence on cloud computing: Evolution, vision, trends and open challenges. *Internet of Things, 8*, 100118.
13. Greco, L., Percannella, G., Ritrovato, P., Tortorella, F., & Vento, M. (2020). Trends in IoT based solutions for health care: Moving AI to the Edge. *Pattern Recognition Letters, 135*, 346–353.
14. Iqbal, S., Kiah, M. L. M., Anuar, N. B., Daghighi, B., Wahab, A. W. A., & Khan, S. (2016). Service delivery models of cloud computing: Security issues and open challenges. *Security and Communication Networks, 9*(17), 4726–4750.
15. Laura Drechsler. (2019, September). *The future of technology and innovation.* Online. Accessed from: https://www.experian.com/blogs/insights/2019/09/the-future-of-technology-and-innovation/
16. Nikitas, A., Michalakopoulou, K., Njoya, E. T., & Karampatzakis, D. (2020). Artificial intelligence, transport and the smart city: Definitions and dimensions of a new mobility era. *Sustainability, 12*(7), 2789.
17. Pandey, R. S., Upadhyay, R., Kumar, M., Singh, P., & Shukla, S. (2020). IoT-based HelpAgeSensor device for senior citizens. In *International conference on innovative computing and communications* (pp. 187–193). Springer.
18. Prathibha, S. R., Hongal, A., & Jyothi, M. P. (2017). IoT based monitoring system in smart agriculture. In *2017 international conference on recent advances in electronics and communication technology (ICRAECT)*. IEEE.
19. Per, L., & Skouby, K. (2016). Complex IoT systems as enablers for smart homes in a smart city vision. *Sensors, 16*, 1840.
20. Philip, A. T. (2018). *Designing and building a hybrid cloud.* O'REILLY. Online. Accessed from https://cdw-prod.adobecqms.net/content/dam/cdw/on-domain-cdw/brands/nutanix/o--reilly-ebook-designing-and-building-a-hybrid-cloud.pdf
21. Radu, L. D. (2020). Disruptive technologies in smart cities: A survey on current trends and challenges. *Smart Cities, 3*(3), 1022–1038.
22. Sethi, P., & Sarangi, S. R. (2017). Internet of things: Architectures, protocols, and applications. *Journal of Electrical and Computer Engineering, 25*, Article ID 9324035. https://doi.org/10.1155/2017/9324035.
23. Shane Wang. (2019). *Artificial intelligence applications in telecommunication clouds.* Online. Accessed from https://01.org/blogs/qwang10/2019/artificial-intelligence-applications-telecom munication-clouds.
24. Siebel, T. M. (2019). *Digital transformation: Survive and thrive in an era of mass extinction.* RosettaBooks.
25. Singh, P., Singh, N., Singh, K. K., & Singh, A. (2021). Diagnosing of disease using machine learning. In *Machine learning and the internet of medical things in healthcare* (pp. 89–111). Academic Press.
26. Singh, K., Arora, G., Singh, P., et al. (2021). IoT-based collection vendor machine (CVM) for E-waste management. *Journal of Reliable Intelligent Environments*. https://doi.org/10.1007/s40860-020-00124-z
27. Singh, P., Singh, N., & Deka, G. C. (2020). Prospects of machine learning with Blockchain in healthcare and agriculture. In *Multidisciplinary functions of Blockchain technology in AI and IoT applications* (pp. 178–208). IGI Global.
28. Singh, P., & Agrawal, R. (2018). A customer centric best connected channel model for heterogeneous and IoT networks. *Journal of Organizational and End User Computing, 30*(4), 32–50.
29. Singh, P., & Singh, N. (2020). Blockchain with IoT and AI: A review of agriculture and healthcare. *International Journal of Applied Evolutionary Computation (IJAEC), 11*(4), 13–27.

30. Singh, P., & Agrawal, R. (2020). An overloading state computation and load sharing mechanism in fog computing. *Journal of Information Technology Research (JITR), 14*(3).
31. Shankar, A., Pandiaraja, P., Sumathi, K., Stephan, T., & Sharma, P. (2020). Privacy preserving E-voting cloud system based on ID based encryption. *Peer-to-Peer Networking and Applications.* https://doi.org/10.1007/s12083-020-00977-4
32. Stephan, T., Al-Turjman, F., Joseph, K. S., Balusamy, B., & Srivastava, S. (2020). Artificial intelligence inspired energy and spectrum aware cluster based routing protocol for cognitive radio sensor networks. *Journal of Parallel and Distributed Computing.* https://doi.org/10.1016/j.jpdc.2020.04.007
33. Talari, S., Shafie-Khah, M., Siano, P., Loia, V., Tommasetti, A., & Catalão, J. P. (2017). A review of smart cities based on the internet of things concept. *Energies, 10*(4), 421.
34. Wan, J., Yang, J., Wang, Z., & Hua, Q. (2018). Artificial intelligence for cloud-assisted smart factory. *IEEE Access, 6*, 55419–55430.
35. Zantalis, F., Koulouras, G., Karabetsos, S., & Kandris, D. (2019). A review of machine learning and IoT in smart transportation. *Future Internet, 11*(4), 94.

Big Data Analytics and Big Data Processing for IOT-Based Sensing Devices

Pawan Kumar Pal, Charu Awasthi, Isha Sehgal, and Prashant Kumar Mishra

1   Introduction

The specialized advancements and brisk converging of radio transmission, arithmetical personal computer (PC) gadgets, and miniature micro-electromechanical system (MEMS) instruments, that is, electromechanical framework instruments, lead to the improvement in Internet of Things (IoT); predictable with the Cisco report [1], the Web's associated measure of items has outstripped the measure of populace in the world. Such objects, which are associated with the Internet, incorporate PCs, cell phones, tablets, Wi-Fi-empowered sensors [29], wearable gadgets, and family apparatuses, and structure IoT, are given in Fig. 1.

According to the reports, the measure of gadgets associated with the Internet has increased usage from 22.9 billion (in 2016) to 50 billion (in 2020) as deduced from Fig. 2. Almost all IoT applications do not just represent considerable authority in checking discrete occasions, but also the information collected by IoT objects is mined. Almost all data assortment devices in the IoT climate are gadgets that have sensors fitted and need custom conventions like the following:

- Message Queuing Telemetry Transport Protocol (MQTT)
- Data dispersion administration (DDS)

P. K. Pal · C. Awasthi (✉) · P. K. Mishra
Department of Computer Science and Engineering, PSIT, Kanpur, Uttar Pradesh, India
e-mail: charu.awasthi@psit.ac.in

I. Sehgal
Department of Information Technology, PSIT College of Engineering,
Kanpur, Uttar Pradesh, India

Fig. 1 Big data sources in IoT

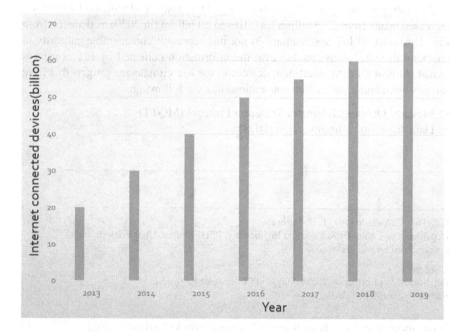

Fig. 2 Internet Connected Devices (Billion)

However, sensors are recommended in virtual businesses, IoT is anticipated to gracefully colossal measure of the information. Information created from the IoT gadgets is frequently incorporated for discovering:

- Potential examination patterns
- Exploration of the effect of specific occasions or choices

Such data are handled using different logical instruments [2].

The advancement of enormous data and IoT is rapidly influencing all territories like innovations and organizations by intensifying unions' preferences (associations) and people. The extension of information made through IoT [] has presumed a genuine element on the enormous data set. To help with comprehension of quite a gigantic idea, data researchers at IBM advocated the four "V's" of massive data: volume, variety, velocity, and veracity [3]:

1.1 Data Volume

Enterprises have ever-developing data, all things considered, effectively accumulating terabytes, even petabytes; for example, changing 350 billion yearly meter readings to raise and foresee utilization. Data sets are regularly enormous to such an extent that they can't fit on one worker and must instead be dispersed between a few stockpiling areas. Data examination programming like Hadoop [34] is made to oblige the need for circulated capacity and conglomeration.

1.2 Data Variety

For time-delicate cycles like getting fakes, huge data must be utilized since it streams into the undertaking. The present advanced data cannot be regularly corralled into conventional structures. Ground-breaking examination programming seems to saddle unstructured data, like pictures and recordings, and blend it in with more direct data streams to gracefully extra bits of knowledge.

1.3 Data Velocity

Big Data [32] comprises a wide range of information—organized and unstructured data like content, sensor data, sound, video, log records, and so on. Presently, data are gathered at an awesome pace of two 0.5 quintillion bytes for each day. From a large number of Web-based media that present on every 5 billion Google search for each day, gathered data are gushing into workers at a formerly exceptional speed.

1.4 Data Veracity

Data veracity alludes to honesty or exactness of a specific arrangement of the information, which has the assessment of information source:

- Is it dependable?
- Or would it not lead the investigators free?

As helpless data quality costs around \$3.1 trillion per annum in the United States, so seeking after veracity is crucial. There are incorporations in it, leading to the following:

- Wipe out duplication
- Limit predisposition
- Cycle data in the manners by which it accumulates for the apparatus or vertical

The overall acknowledgment of IoT has made huge data examinations fascinating due to the preparation as well as an assortment of data by a particular sensor encompassed by the IoT biological system. The International Data Corporation (IDC) article showed that the gigantic data commercial center led to an increase of over US\$125 billion by the year 2019 [4]. Figure 3 shows the strategy for information assortment, checking, and data investigation [5].

Large data examination enables data diggers and researchers to investigate gigantic measures of unstructured data that will be used for utilizing customary apparatuses [6] along with huge data examination that woks to immediately remove educated data-utilizing and data-preparing procedures that help in:

- Creating expectations
- Distinguishing ongoing patterns
- Finding concealed data
- Settling on the choices [7]

Procedures in data preparation are broadly sent for both issue explicit techniques and summed up data investigation. As needs be, measurable and AI techniques are used. IoT data are unique in relation to ordinary huge data gathered by means of frameworks as far as attributes due to different sensors as well as articles required during data assortment. It incorporates:

- Heterogeneity
- Commotion
- Assortment
- Fast development

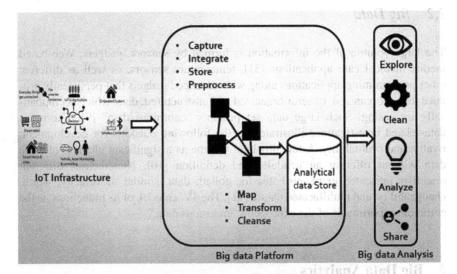

Fig. 3 Strategy for information assortment, checking, and data investigation

2 Outline of IoT Along with Big Data

2.1 Internet of Things (IoT)

IoT bids a stage for sensors as well as gadgets to talk flawlessly in a reasonable climate and empowers data sharing across stages in an advantageous way. The ongoing variation in different remote innovations places IoT on the grounds that the following progressive innovation by making the most of the total open doors offered by the Web's novelty. IoT has witnessed its ongoing appropriation in the shrewd urban areas and interest in creating keen frameworks, similar to office, retail, shrewd agribusiness, keen water, keen transportation, and smart medical services [30], and great energy [8].

Numerous specialized gadgets in the IoT worldview are implanted into sensor gadgets in the world. Data putting away contraptions sense measurements and move these data utilizing installed correspondence machines. The range of gadgets along with articles is interlocked through a spread of the correspondence arrangements, such as Bluetooth, Wi-Fi, Zigbee, and Global System for Mobile Communication (GSM). Such specialized gadgets transmit data and get orders from the distantly controlled gadgets. These orders license direct reconciliation with the actual world through PC-created frameworks to upgrade expectations for everyday comforts.

2.2 Big Data

The large volume of the information is formed by sensors, gadgets, Web-based media, medical care applications [31], temperature sensors, as well as different other programming applications along with advanced gadgets that persistently produce lots of organized, or semi-organized and unstructured, data, which is emphatically increasing. Such large data set prompts "enormous data" [9]. Enormous data-related investigations illustrate that the following outskirts for development, rivalry, and profitability. A few proclaim volume as a significant element of large data without offering an unadulterated definition [10]. Nonetheless, different scientists presented extra attributes for goliath data, similar to veracity, worth, changeability, and multifaceted nature [11]. The 4V's model, or its inductions, is the commonest portrayals of the expression "enormous data."

3   Big Data Analytics

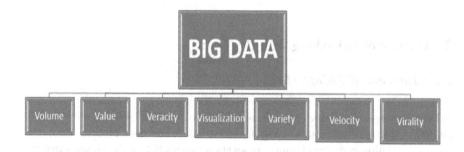

It includes cycles of exploring through a database, mining, and breaking down data dedicated for improving organization execution [12]. It is the strategy for analyzing huge data sets that contain a spread of information types [13] to discover concealed examples like masked relationships, market patterns, client preferences, and other useful business data. Most of the target of Big Data analytics is to aid business relationships to get improved comprehension of information and settle on skillful choices. Big Data analytics requires advancements and apparatuses that will change an outsized measure of "organized, unstructured, and semi-organized data" in the form of a more reasonable data and metadata design for logical cycles. These scientific instruments' calculations should find examples, patterns, and connections throughout a spread of timelines in the data [14]. These instruments envision the discoveries in tables, diagrams, and spatial graphs for effective choosing after investigating the data. Subsequently, Big Data examination might be a genuine test for a few applications because of data intricacy and the adaptableness of basic calculations that support such cycles [15].

4 Variety of Data Types

The time associated with the origin of Big Data has delivered spread of different data sets from various sources in a few areas. These data sets include various modes, every one of which features a particular depiction, dispersion, scale, and thickness.

4.1 Network Data (Online)

One of the most focal points related to organization's Big Data is online informal community (online social networks [OSNs]) data, such as Facebook [16] and Second Life [17]. Center has expanded with headways in data investigation. Numerous examinations are performed with online informal communities utilizing information about delegate attributes at the full-scale level, for instance, little world highlights. Nonetheless, factors for the highlights of potential miniature cycles aren't very much spoken to in these investigations.

4.2 Mobile and IoT Data

The most common pattern of organization's Big Data is the investigation of portable and IoT data. With the occasion of 5G innovation, joined portable organizations have brought about critical upgrades in machine-to-machine interchanges execution. Joined versatile networks uncover relinquished reach groups in cellulite organizations, as Long-Term Evolution-Advanced, by utilizing intellectual radio innovation.

4.3 Geography Data

One examination tends to the gauntlets of significant kinds of innovation for three-dimensional (3D) association, and volume-delivering innovation upheld graphics processing unit (GPU) innovation. This work investigates visual programming for the hydrological climate upheld data direction. Also, it produces sea plans and forms planning of surfaces, component field planning, and dynamic recreation of the predominant field [18]. To bring present highlights up in space and accomplish continuous overhauling of an outsized measure of hydrological climate data, the examination builds hubs on the spot to control math to acknowledge dynamic planning of high properties.

4.4   Spatial–Temporal Data

Data are characterized in numerous classes upheld highlights and contrasts. Since the distinctions in data decide the accomplishment of the examination, they assume a significant job. Various highlights likewise are applied to search for a comparable highlight. Data in huge databases are regularly recovered by data preparing. In the instance of time-evolving data, when time becomes associated, the information is mined regarding both realities. The investigation of information mining as far as both existences has impacted the investigation of cell phones' investigation [19]. For the most part, spatial Big Data is engaged with vector data and raster data for the most part, and spatial Big Data is engaged with vector data, raster data, and organization data. The issue with utilizing databases from the disposition of room is that there are numerous obstructions at the security level.

4.5   Streaming as well as Real-Time Data

Amid the expansion for Internet Web-based features, network Big Data has advanced from spatial–worldly data to constant spatial–transient data. Gathering overviews for the most part needs consistent measurements' study owing to the constant reclamation of reports and insights over huge sum of data streams. The age of Big Data has started, and far of the information are wont to investigate the dangers of a spread of business applications. There are innovative preliminaries in the assortment of Big Data during a complex indoor modern climate. Indoor remote sensor organization (wireless sensor network [WSN]) innovation can conquer such limitations by getting together with Big Data from source hubs.

4.6   Visual Data

Within the period of Big Data, truly expanding measures of picture data should have presented critical difficulties to current picture investigation and recovery. It is vital to record pictures productively and adequately with semantic catchphrases, especially when gone up against the on the Web's quick developing properties.

4.7   Data's Associated Challenges

Every extraordinary data space increases difficulty, which, appropriately tended to, may crucially affect cutting-edge Big Data frameworks. In the primary spot, online organization data remain anticipating better models, with an expanded help from sociologists. Versatile data and the IoT, which produce a lot of information, would appreciate the selection of a gigantic data foundation prepared to store and handle data in current IoT frameworks.

5   Role of Big Data in IoT

IoT acts as a significant hot spot for data that companies are getting for examination. That is the reason for the emergence of IoT in Big Data. Analytics of Big Data is arising as a vital aspect for dissecting IoT created data from "associated gadgets," which assist the activity to upgrade required.

The function of Big Data in IoT is to handle an outsized measure of information reliably and putting away them utilizing diverse-capacity technologies:

IoT-based Big Data preparation follows four successive steps:

- In the framework of Big Data, IoT gadgets create an outsized measure of unstructured data, which have three variables: volume, speed, and assortment. These variables are generally called as 3V's.

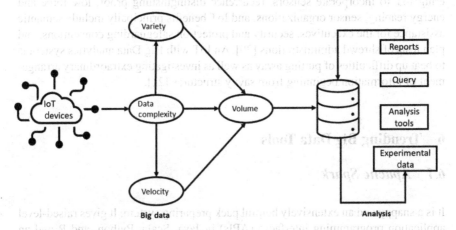

- Within a common disseminated database (the Big Data framework), the enormous measures of information are put away in Big Data documents.
- The utilization of logical apparatuses like Hadoop MapReduce or Spark for analyzing the put-away IoT Big Data.
- Producing the reports of examined data.

Since in IoT the unstructured data are gathered by means of the Web, subsequently, Big Data for the trap of things needs extremely quick investigation with more questions to acknowledge fast bits of knowledge from data to shape speedy choices. Subsequently, the need for Big Data in IoT is convincing.

5.1   Recent Advances in IoT-Based Big Data and Analytics

Analytics of Big Data might be a rapidly growing research territory covering the areas of figuring, data on the board, and has become a universal term in agreement and taking care of multifaceted issues in a few punitive fields like applied math, medication, computational science, medical care, interpersonal organizations, money, business, government, schooling, transportation, and telecommunication [20]. The utility of Big Data is found to a great extent inside the region of the Internet of Things (IoT). People have been increasingly moving to urban areas in recent years. By 2030, more than 60% of the world's population will be living in urban areas. Cities are Big Data resources because they include millions of people, technical devices, and cars, all of which produce data on a regular basis. Green Internet of Things (G-IoT), as implemented by Chithaluru et al. [27] for smart cities, plays one of the most important roles in the development of a green and safe-living environment. Big Data analytics is critical for gaining useful information from large amounts of data provided by the Internet of Things.

Big Data is utilized to make IoT structures that incorporate things-driven, data-driven, administration-driven design and cloud-based IoT. Advancements empowering IoT to incorporate sensors, recurrence distinguishing proof, low force and energy reaping, sensor organizations, and IoT benefits principally include semantic assistance for the executives, security and protection safeguarding conventions, and plan tests of shrewd administrations [20]. An IoT with Big Data analytics system is to beat up difficulties of putting away as well as investigating extraordinary arrangement of information beginning from savvy structures [21].

6   Trending Big Data Tools

6.1   Apache Spark

It is a snappy and an extensively helpful pack preparing system. It gives raised-level application programming interfaces (APIs) in Java, Scala, Python, and R and an improved engine that endorses typical implementation diagrams [22]. It also endorses a powerful strategy of more critical degree instruments, including Spark SQL for SQL and coordinated data taking care of MLlib for AI, GraphX for chart planning, and Spark Streaming. Shimmer is proposed to include a wide extent of outstanding weights, for example, bunch applications, iterative figuring, savvy questions, and streaming. Besides supporting all these remarkable jobs needing to be done in a specific structure, it diminishes the organization's weight of keeping up discrete gadgets.

It has the following characteristics.

(a) *Speed*—It helps run an application in the Hadoop pack, up to different occasions snappier in memory and on numerous occasions faster while passing on circle. This is done by diminishing amount of scrutinizing/forming exercises to plate. It empowers the center getting ready data in memory.

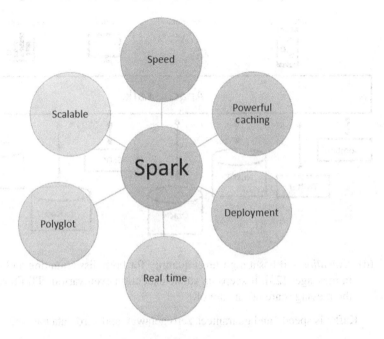

(b) *Supports multiple dialects*—Spark gives specific APIs in Java, Scala, or Python. Subsequently, we can make functions in distinct lingos. Streak prepares 80 raised level directors for wise addressing.
(c) *Advanced analytics*—It not simply sponsorships "Guide" and "lessen". It encourages SQL questions, streaming data, ML, and graph algorithms.

6.2 Apache Kafka

It is an organization scattered capacity streaming stage fit for managing trillions of capacities daily. From the beginning envisioned as an advising line, Kafka relies upon an impression of an appropriated submit log. Since being made and freely delivered by LinkedIn in 2011, Kafka has promptly evolved from advising line to a certain capacity streaming stage.

The following are several focal points of Kafka:

(a) *Unwavering quality*—Kafka is appropriated, separated, rehashed, and transformation to inside disappointment.

(b) *Versatility*—Kafka's illuminating structure scales viably without individual time.

(c) *Strength*— It uses dispersed submit log which requires communications suffers on hover as speedy as it could be normal the situation being.

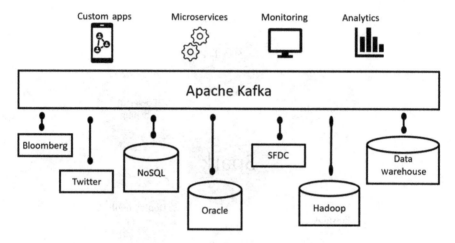

(d) *Execution*—It has a high-level quantity for both disseminating and purchasing in messages [23]. It keeps up stable execution even various TB (Tera Bytes) of the messages are taken care of.

Kafka is speedy and guarantees zero getaway and zero data hardship.

6.3   Flink

Apache Flink is an open-source architecture for batch and stream processing, and it is powerful and extremely accurate in data ingestion. Flink is capable of running under all basic and necessary conditions and measures in-memory speed at any scale. By planning and effective distribution of load, it gives a high throughput and low latency for processing streams with its fast streaming engine. It works on both bounded and unbounded data sets. Unbounded streams are processed continuously, and bounded streams have defined start and end sessions. The real-time data as they are sensed can be processed and later stored into a file system for further application. Flink is capable of doing interactive processing, batch processing, in-memory processing, real-time stream processing, iterative processing, and graph processing. Its dataflow applications can be written in Java, Python, and SQL and can be configured in cloud atmosphere also. It supports to Amazon Kinesis, Apache Cassandra, HDFS, and Apache Kafka.

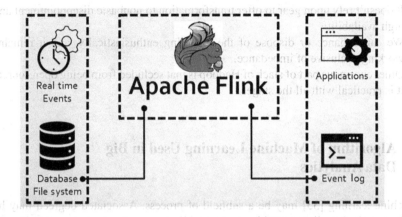

6.4   Hadoop

The Apache's Hadoop programming collection is the structure that considers the dispersed treatment of immense enlightening files across gatherings of personal computers (PCs) using essential programming models. It is proposed to scale up from single laborers to many machines, each offering neighborhood estimation and limit. Instead of relying upon hardware to pass on high openness, the library itself is expected to recognize and manage discontents for the product layer, so passing on an incredibly available help at top of a lot of PCs, all of which may be slanted to disillusionments.

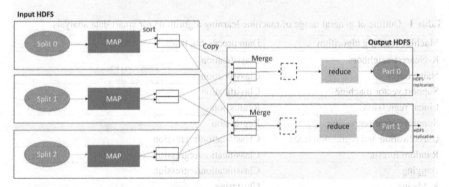

The following are the central purposes of using Hadoop:

- Its assembly permits the client to style as well as test passed-on systems speedily. It is beneficial, and it customizes appropriate data and works over the machines and, in this manner, utilizes the essential parallelism of the central processing unit (CPU) communities.

- It doesn't rely upon gear to offer transformation to nonbasic disappointment and high availability.
- We can enhance or dispose of the gathering enthusiastically, and it remains working exclusive of impedance.
- Other enormous part of slack of Hadoop is that secluded from being open source; it is practical with all the stages.

7 Algorithm of Machine Learning Used in Big Data Analytics

Machine learning [33] may be a subfield of process. Associated degreed may be such an (AI) that allows machine to find while not unequivocal programming. AI degenerated from style recognition and machine learning hypothesis. Here, some basic ideas of AI are examined to boot in lightweight of the fact that the as usually as potential applied AI calculations for sensible information examination. A learning calculation accepts a gathering of tests as associate degree data named as preparation set by and enormous to it, there exist three primary categories of learning: administered, solo, and fortification.

In managed learning, preparation sets contain tests of information vectors nearby their related fitting objective vectors, likewise alluded to as marks. In the unaided type of learning, no marks are needed for the preparation set. Fortification type of learning manages matter of learning the satisfactory activity or grouping of moves to be created for associate degree-offered circumstance to spice up the result.

They applied AI calculations for sensible information investigation, and IoT use cases are presented in Tables 1 and 2.

Table 1 Outline of general usage of machine learning algorithms for smart data analysis

Machine learning algorithm	Data processing tasks
K-Nearest neighbors	Classification
Naive bayes	Classification
Support vector machine	Classification
Linear regression	Regression
Support vector regression	Regression
Classification and regression trees	Classification/regression
Random forests	Classification/regression
Bagging	Classification/regression
K-Means	Clustering
Density-based spatial	Clustering
Clustering of applications with noise	
Principal component analysis	Feature extraction
Canonical correlation analysis	Feature extraction
Feed forward neural network	Regression/classification/clustering/feature extraction
One-class support vector machines	Anomaly detection

Table 2 Overview of applying machine learning algorithms to the Internet of Things use cases

Machine learning algorithm	IoT, smart city use cases	Metric to optimize
Classification	Smart traffic	Traffic prediction, increase data abbreviation
Clustering	Smart traffic, smart health	Traffic prediction, increase data abbreviation
Anomaly detection	Smart traffic, smart environment	Traffic prediction, increase data abbreviation, finding anomalies in power dataset
Support vector regression	Smart weather prediction	Forecasting
Linear regression	Economics, market analysis, energy usage	Real time prediction, reducing amount of data
Classification and regression trees	Smart citizens	Real time prediction, passengers travel pattern
Support vector machine	All use cases	Classify data, real time prediction
K-Nearest neighbors	Smart citizen	Passengers' travel pattern, efficiency of the learned metric
Naive bayes	Smart agriculture. Smart citizen	Food safety, passengers travel pattern, estimate the numbers of nodes
K-Means	Smart city, smart home, smart citizen, controlling air and traffic	Outlier detection, fraud detection, analyze small data set, forecasting energy consumption, passengers travel pattern, stream data analyze
Density-based clustering	Smart citizen	Labeling data, fraud detection, passengers travel pattern
Feed forward neural network	Smart health	Reducing energy consumption, forecast the states of elements, overcome the redundant data and information
Principal component analysis	Monitoring public places	Fault detection
Canonical correlation analysis	Monitoring public places	Fault detection
One-class support vector machines	Smart human activity control	Fraud detection. Emerging anomalies in the data

7.1 K-Nearest Neighbors (KNN)

In KNN, the target is to order a substitution assumed inconspicuous information by viewing the K-given information focuses within the preparation set nearest to it within the information highlight house. Thus, to go and look out the KKN of the new information, we would value more highly to utilize a distance metric, as an example, geometer distance, L∞ normal, point, Mahalanobis distance, or playacting distance. One restriction of KNN is that it needs golf shot away the complete getting-ready set that makes KNN unclimbable to very large information sets.

Working of KNN: The KNN operation can be processed as per the below calculation:

Step-1: choose the amount K of the neighbors.

Step-2: Calculate the geometer distance of K range of neighbors.

Step-3: Take the K nearest neighbors consistent with the determined geometer distance.

Step-4: Among these k neighbors, check the amount of the info focuses on each classification.

Step-5: Assign the new information focuses to it category that the amount of the neighbor is greatest.

Step-6: Our model is ready.

7.2 Naive Bayes

It is a characterization technique upheld theorem, which has supposition of autonomy among indicators. Generally, Naive Bayes categorization expects that the presence of a selected component during a class is inconsequential to the presence of another part. Guileless Bayes model is something however tough to fabricate and particularly valuable for exceptionally immense information sets. aboard straightforwardness, Naive Bayes is understood to beat even deeply fashionable arrangement methods.

Bayes hypothesis provides technique of computing back chance $P(c|z)$ from $P(c)$, $P(z)$, and $P(z|c)$. Take a gander at the condition shown below:

$$P(c \mid z) = \left(P(z \mid c)P(c) / P(z) \right)$$

$$P(c \mid Z) = P(z_1 \mid c) * P(z_2 \mid c) * \ldots \ldots * P(z_n \mid c) * P(c)$$

where:

$P(c|z)$ is the posterior probability of class (c, target) give predictor (z, attribute).

$P(c)$ represents class prior probability.

$P(z|c)$ represents the probability of class predictor.

$P(z)$ gives the prior probability predictor.

7.3 Support Vector Machine

Classical support vector machines (SVMs) are nonprobabilistic, parallel classifiers that will dig up the analytic hyperplane that isolates the two categories of the preparation set with the foremost extreme edge. At that time, the substitution's anticipated name, hid information is chosen upheld that facet of the hyperplane it falls

[24]. To start with, we tend to examine the straight SVM that discovers a hyperplane which is a right way capability of knowledge variable.

7.4   Linear Regression

Linear relapse is one of the best and most well-liked machine learning calculations. It is a mathematical technique that is used for logical examination. Straight relapse makes forecasts for consistent/genuine or numeric factors, for example, deals, pay, age, item cost, and so on.

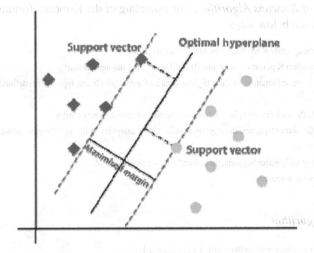

Direct relapse calculation shows a straight association between a dependent loved one (x) and a minimum of one relative (x) factors, later on known as direct relapse. Since direct relapse shows the straight relationship, which means it discovers, the dependent relative variable's estimation is moving scrutiny to the estimation of the free issue.

The direct relapse model offers a slanted line chatting with the association between the factors.

In fact, we can define a straight relapse as follows:

$$x = b_0 + b_1 z + \varepsilon$$

where:

x = Dependent variable (target variable)
z = Independent variable (predictor variable)
b_0 = Intercept of the line
b_1 = Linear regression coefficient (scale factor for input value)
x and z are the input data set values

7.5 K-means Algorithm

It is an associate with Big Data reiterative formula that separates the untagged data set into k distinct clusters in such a way that each data set slot in just one cluster that has connected assets. It is a centroid-based formula, where each cluster relates to a center of mass. The most important target of this procedure is to cut back the addition of areas between the information purpose and their consequent clusters. It performs primarily two tasks that is decide the simplest worth for K center points or centroids by the associate in Big Data reiterative facet and allocates every information to its nearest k-center. Those information points that square measure just about the k-center, produce a cluster.

Working of K-means Algorithm: The operating of the K-means formula is delineated within the below steps:

Step-1: choose the quantity K to choose the quantity of clusters.

Step-2: choose random K points or centroids. (It is different from the input dataset).

Step-3: Assign every information to their highest centre of mass, which can type the predefined K clusters.

Step-4: Calculate the variance and place a brand-new centre of mass of every cluster.

Step-5: Repeat the third steps, which suggests transfer every datapoint to the new highest centre of mass of every cluster.

Step-6: If any duty assignment happens, then attend step-4 else attend end.

Step-7: The model is prepared.

K-means Algorithm

Input: K, and unlabeled data set $\{x_1, \ldots, x_N\}$.

Output: Cluster centers $\{s_k\}$ and the assignment of the data points $\{\pi_{nk}\}$.

Randomly initialize $\{s_k\}$.

repeat

    for $n := 1$ *to* N **do**

        for $k := 1$ *to* K **do**

            if $k = \arg\min_i \|s_i - x_i\|^2$ **then**

                $\pi_{nk} := 1$

            else

                $\pi_{nk} := 0$

            end

        end

    end

    for $k := 1$ *to* K **do**

        $s_k := \dfrac{\sum_{n=1}^{N} x_n \pi_{nk}}{\sum_{n=1}^{N} \pi_{nk}}$

    end

until $\{\pi_{nk}\}$ *or* $\{s_k\}$ *don't change;*

7.6   Principal Component Analysis (PCA)

Principal component analysis is an unsupervised learning formula that is applied to the spatiality cut in mil. It is an arithmetic method that adapts the reflections of coupled characteristics into a group of linearly unrelated characteristics with the assistance of orthogonal revolution. These new altered characteristics of square measure are known as the principal parts. It is one of the quality tools that are applied for experimental information analysis and prognosticative modeling.

Working of Principal Component Analysis: The operating of the principal part analysis is delineated within the below steps:

Step 1. Obtaining the data set
Step 2. Signifying information into a structure
Step 3. Normalizing the information
Step 4. Analyzing the variance of Z
Step 5. Analyzing the Eigenvalues and Eigenvectors
Step 6. Sorting the Eigenvectors
Step 7. Analyzing the new characteristics or principal element
Step 8. Eliminate less or irrelevant characteristics from the new data set

PCA Algorithm

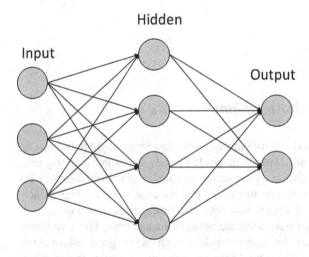

7.7   Neural Network

A neural network may be a kind of approach that models the system as the human brain, forming a man-made neural network (ANN) via associating in Big Data formula that permits the pc to be instructed by group actions and new information. Their square measure variant AI algorithms lately, neural networks square measure capable of working what has been characterized deep learning. Whereas the easy a part of the brain is that the somatic cell, the crucial developing block of Associate in Big Data ANN may be a perceptron that achieves a straightforward signal process. These square measures are then connected into an outsized mesh network.

Input: L, and input vectors of an unlabeled or labeled data set $\{x_1, \ldots, x_N\}$.

Output: The projected data set $\{z_1, \ldots, z_N\}$, and basis vectors $\{w_j\}$ which form
the principal subspace.

$\bar{x} := \frac{1}{N} \sum_n x_n$

$S := \frac{1}{N} \sum_n (x_n - \bar{x})(x_n - \bar{x})^T$

$\{w_j\} :=$ the L eigenvectors of S corresponding to the L largest eigenvalues.

for $n := 1$ *to* N **do**

   for $j := 1$ *to* L **do**
      $z_{nj} := (x_n - \bar{x})^T w_j$
   end

end

7.8   IoT's Data Analytic Algorithms

Reminiscent of standard good statistics and analytic algorithms ought to be capable of handling massive information, which means IoT needs algorithms that may analyze information that comes from an expansion of sources in period. Several tries square measure created to subsume this issue. For instance, deep learning principles, that measure a type of neural networks integrating evolution will achieve higher accuracy rate and they have adequate information and time. The deep learning process will effortlessly be manipulated through noisy good information. Moreover, the algorithm based on neural network lacks interpretation, that is, information scientists cannot perceive the model results' motivations. At intervals identical manner, semi-supervised algorithms, that models a tiny low quantity labeled information along with associate in Big Data outsize quantity of untagged information, will assist IoT information analysis.

8 Types of Technologies of Big Data

Big Data is a technology that [35] is mainly classified into two types:
1. Big Data technology for data ingestion and operation
2. Big Data technology for analytical and computational requirements

8.1 Operational Big Data Technology

It is applicable for generation and collection of operational data used by different applications every day and applying different operations to identify meaningful information that will be utilized for further processing. This can be collected from Web transactions, sensor networks, social media, or the data from a selected organization; then, we'll even believe this to be a form of information that is used to require care of the analytical massive information technologies (ITs).

Few cases of operational massive information technologies are shown below:

SOCIAL MEDIA MINING

- Booking online ticket, for train, flight, movie, bus, etc.
- Online searching about any activity, event, and news related to people and different organizations.
- Collection of data from different social media platforms like Instagram, Amazon, Facebook, etc. We can generate a large volume of categorical data based on activities of users on theses platforms.
- The worker information of any transnational organization.

8.2 Analytical Big Data Technology

It resembles the intense adaptation of huge information technologies. It is somewhat complex than the operational massive information. To place it simply analytical monumental information is required for the representation of important business selections by functional big data as shown in the image.

Few cases of analytical Big Data technologies are as follows:

- Sentimental analysis
- Stock marketing
- Crowd management
- User-specific customized services
- Space missions
- Weather prediction
- Medical fields

9 Trending Technologies of Big Data

They are distributed into four areas that are categorized as shown below:

- Data collection and storage
- Data retrieving and mining
- Data computation, processing, and analytics
- Data visualization

9.1 Data Storage

9.1.1 MongoDB

The NoSQL document databases like MongoDB supply an instantaneous possibility in distinction to the unbending blueprint used in relative databases. It allows MongoDB to supply flexibly a relatable assortment of data types all over quantity and across distributed architectures. MySQL, Microsoft SQL Server, and MS Access business applications are now be implemented using MongoDB due to variety of dimention of data.

9.1.2 RainStor

RainStor may be a product organization that designed up a management system of an analogous name meant to achieve and analyze massive information for big ventures. It utilizes deduplication techniques to plan out the method toward swing-away loads of data for reference. Barclays and Credit Suisse measure victimization RainStor.

9.2 Data Mining

9.2.1 Presto

It is an open source for executing interactive analytic queries; it uses SQL query engine alongside data wellsprings of every sizes going from gigabytes to petabytes. It allows addressing data in Hive, Cassandra, Relational Databases, and Proprietary Data Stores. It is made in the year 2013 by Apache Foundation. Facebook, Netflix, and Airbnb are a couple of associations that are using Presto for data mining.

9.2.2 Rapid Miner

It is a centralized arrangement that includes an incredible and powerful graphical user interface (GUI) that empowers clients to create, deliver, and keep up predictive analytics. It permits making advanced workflows and scripting support in a few dialects. Domino's, Infocus, BCG, and Slalom are organizations that are using Rapid Miner.

9.2.3 Elasticsearch

It is a search engine reliant on the Lucene Library. It gives a:

- Distributed
- Multitenant-capable
- Full-text search engine with a HTTP Web interface and without pattern JSON chronicles.

Accenture, LinkedIn, Netflix, and Stack Flood are using Elasticsearch for data mining.

9.3   Data Analytics

9.3.1   Kafka

It is a distributed streaming stage. The streaming stage has three key abilities that are according to the accompanying:

- Distributer
- Endorser
- Shopper

This resembles:

- A message queue or
- An enterprise messaging system

It is created in 2011 by Apache Foundation. Netflix, Yahoo, LinkedIn, Twitter, and Spotify are a few organizations that are using Kafka.

9.3.2   Spark

It gives in-memory computing capacities in order to pass on speed. A generalized execution model helps a wide range of Java, Scala, and Python APIs for effortlessness for progress. Prophet, Amazon, Cisco, and Verizon are a few organizations that are using Spark.

9.4 Data Visualization

9.4.1 Tableau

It is a potent and the quickest software creating data visualization instrument operated in the business intelligence industry. Data assessment is incredibly speedy with Tableau, and the visualizations made are as dashboards and worksheets.

9.4.2 Plotly

It is generally used to make graphs snappier and more successful. Programming interface libraries are available for:

- Python
- R
- MATLAB
- Node.js
- Julia
- Arduino
- REST API. Plotly can be used to design interactive graphs with Jupyter notebook.

10 Real-World Applications of Big Data

Here are few examples of actual applications of Big Data:

10.1 eBay (Ecommerce)

a. Recommendation engines
b. Ad targeting
c. Search quality
d. Abuse and click fraud detection

10.2 JP Morgan Chase (Banking and Finance)

a. Modeling true risk
b. Threat analysis
c. Fraud detection

d. Trade surveillance
e. Credit scoring and analysis

11 Privacy and Data Security

Due to the extent of Big Data, well-being and security insurance might be an urgent issue [25]. There could likewise be dangers of protection infringement at each stage of progression. There are various procedures for classification shield, for example, encryption. The acknowledgment of Big Data relies upon a whole comprehension of the security issues inside the framework. Security might be another worry, and this proposal predominantly presents the concept of protection utilizing new issues and spotlights on proficiency and protection assurance. This examination centers around the structure of Big Data analytics, exhibiting protection assurance needs; moreover, it clarifies the security insurance coined closeness arrangement in data handling and prerequisites. The World Data Corporation overview indicated that, in 2011, 1.8 trillion gigabytes of information was made, and replicated, the sum of which is copied like clockwork. In the coming decade, the whole measure of information focus oversaw data will be multiple times bigger; notwithstanding, proficient IT staff will be able to handle it. Regular instruments aren't prepared to measure and influence the information contained during this measure of information, nor would they be able to guarantee security. Centered on Big Data, Bhardwaj et al. [] suggested a framework for recognizing cyber-targeted attacks and detecting a suspicious activity.

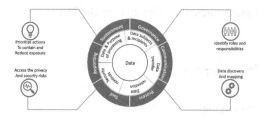

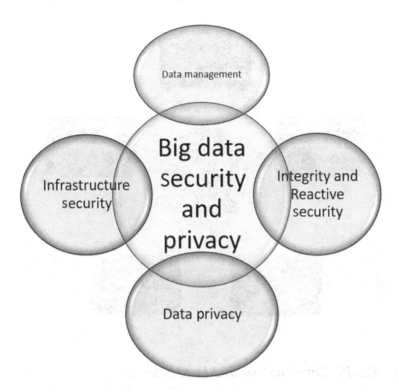

12   Applications of Big Data

As data and collaborations are produced in such a human conduct, Big Data is utilized in many parts of life. The Big Data progressively grants both exploration and mechanical fields, similar to medical care, monetary administrations, and business proposals. Big Data is utilized fundamentally in order to foresee certain ideas, similar to stock costs.

12.1   Behavior Prediction

Most of the organization's Big Data forecasts are upheld data from OSNs. Big Data is utilized in expectations upheld positioned data, similar to decisions, vehicle execution, and different territories in business and legislative issues

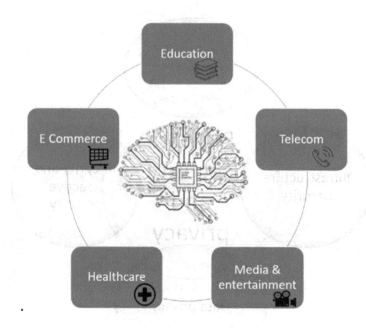

12.2   Health-Care Analysis and Data Storage

Health care might be a multidimensional framework setup with prime focus on the following:

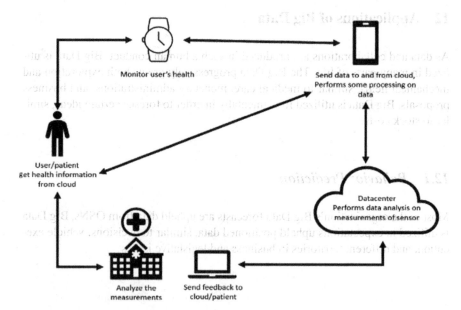

- Avoidance
- Determination
- Treatment of well-being, which are related issues in populace. The principal segments of a medical care framework are the doctors or attendants, emergency clinics for transmitting medications, and other finding or therapy advancements, along with a financing establishment supporting the last two. The well-being experts have a place with differed well-being areas like dentistry, medication, maternity care, Big Data, brain science, physiotherapy, and heaps of others. Medical care is needed at a few levels depending on the earnestness of circumstance.

12.3 Content Recommendation

Recommender frameworks are one among the preeminent, normal, and basically reasonable uses of Big Data. The first realized application is Amazon's suggestion motor, which furnishes clients with an altered site once they visit Amazon.com. Nonetheless, online business organizations aren't the sole ones that utilize proposal motors to impact clients to look for extra items. There are use cases in amusement, gaming, training, publicizing, home stylistic layout, and various ventures. The strategies have different capacities, from proposing music and occasions to furniture and dating profiles. Numerous overall known industry pioneers spare billions of dollars and have connection with a few times more clients be tackling the office of recommender frameworks. Henceforth, Netflix says they spare $1 billion yearly, and around 75% of substance clients traverse proposals.

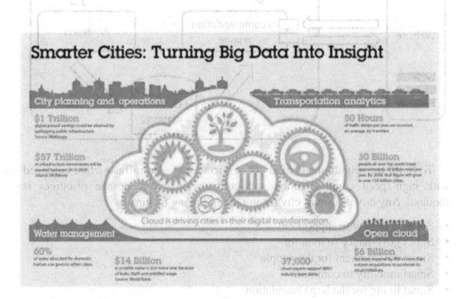

12.4 Smart City

Smart cities are the need of society in all aspects to fulfill the requirements and needs of growing urban population. Nowadays, smart cities are in existence because of the technology improvement and latest innovations. Providing quality services to such a voluminous city is only possible with technology integration with all services required within the city. Massive information is growing speedily, presently at a projected rate of 40, the enlargement in the quantity of worldwide information generated every year, versus solely 5, the enlargement in world IT payment. Inclusion of information and communication technology for different requirements of a city such as transportation, building construction, energy, health care, education, and safety improved the living standard of people and also made it possible to serve the growing population of cities. Big Data in health care refers to vast amounts of data generated by the use of digital technology that capture medical records and aid in the management of hospital results, which would otherwise be too broad and complicated for conventional technologies. At any of these stages, health-care providers are in charge of various types of records, including the patient's medical history (including diagnoses and prescriptions), medical and therapeutic data, such as the one proposed by Punitha et al. [29], and other private or confidential medical data.

Recommendation System

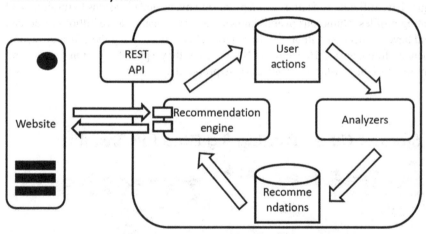

To fulfill the requirements of above services, advanced infrastructure equipped with high-speed communication and computational and storage resources is required. Any city is a smart city having the following features:

* Smart economy to support city
* Smart governance to monitor
* Smart environment for the people
* Smart mobility to commute
* Smart living for the large population

- Smart people and operations to utilize all facilities and services provided by smart cities

Development of smart cities is based on technology factors, institutional factors, and human factors. Huge migration of people from villages to urban areas all over the world has resulted in development of smarter cities.

13   Future Work

The value inside the updates in massive information lies in the information sorting, analysis of algorithms, and new products. In the last few years, enlargement into massive information has been closely linked to mobile and devices. The growing quality of online things has conjointly produced new styles of massive information, and numerous networking styles enable the interconnection of variable networking information. The important good applications for large information have:

- Integrated media
- Communications
- Social networking
- Sensors

The prospects for information assortment are obtaining vital as well as solely helpful information being gathered to unravel pressing problems. Regarding huge information, the latest facilities have provided larger accessibility and quality, permitting additional versatile and efficient procedures for terminal devices and material assortment. The worth of deed material has been lowered, benefiting period assortment and process of huge information, ever-changing the affiliation in between decision-making information, and the possibility of obtaining the correct information model from the data. Information technology has been widely accepted as artificial network optimization tool that supports the following:

- Complete innovation
- Information assortment
- Updates and identification
- Correspondence which can become additional machine controlled

   Many nations have begun to implement:

- New information security technology
- New information protection laws

Huge information security is becoming harsher. As far as information security is concerned, the public is more involved in protecting personal privacy [28] than trade secrets.

14 Conclusion

In a current scenario of smart living and fulfilling the requirements of huge population, applications of IoT devices are growing. This growing network of sensor devices generates huge amount of data that need to be processed to identify information required for IOT-based intelligent application. Big Data analytics plays a significant role in generating information from raw data generated by IoT devices. Different tools and technologies of Big Data analytics are capable of classifying and processing this Big Data. Different tools designed with mathematical and analytical principles can classify and process such a huge volume of data and provide application-specific results faster and in a timely manner. This work highlighted all such Big Data analytics tools in an information and a knowledge context.

References

1. https://dl.papergram.ir/mobileapp1/bigdata/54/e309.pdf
2. Yaqoob, I., Hashem, I. A. T., Gani, A., Mokhtar, S., Ahmed, E., Anuar, N. B., & Vasilakos, A. V. (2016). Big data: From beginning to future. *International Journal of Information Management, 36*(6), 1231–1247.
3. https://www.soracom.io/blog/what-is-the-relationship-between-iot-and-big-data/
4. Gantz, J., & Reinsel, D. (2011). Extracting value from chaos. *IDC iView, 1142*, 1–12.
5. http://www.businessinsider.com/how-the-internet-ofthings-market-will-grow-2014-10
6. Golchha, N. (2015). Big Data–The information revolution. *IJAR, 1*(12), 791–794.
7. Tsai, C.-W., et al. (2015). Big data analytics: A survey. *Journal of Big Data, 2*(1), 1–32.
8. Al Nuaimi, E., et al. (2015). Applications of big data to smart cities. *Journal of Internet Services and Applications, 6*, 25.
9. Kambatla, K., et al. (2014). Trends in big data analytics. *Journal of Parallel and Distributed Computing, 74*(7), 2561–2573.
10. Borkar, V., Carey, M. J., & Li, A. C. Inside "big data management": Ogres, onions, or parfaits? In *Proceedings of the 15th international conference on extending database technology, EDBT '12, 2012* (pp. 3–14). ACM.
11. Gani, A., et al. (2016). A survey on indexing techniques for big data: Taxonomy and performance evaluation. *Knowledge and Information Systems, 46*(2), 241–284.
12. Kwon, O., Lee, N., & Shin, B. (2014). Data quality management, data usage experience and acquisition intention of big data analytics. *International Journal of Information Management, 34*(3), 387–394.
13. Mital, R., Coughlin, J., & Canaday, M. (2015). Using big data technologies and analytics to predict sensor anomalies. In S. Ryan (Ed.), *Proceedings of the advanced Maui optical and space surveillance technologies conference, held in Wailea, Maui, Hawaii, September 15–18, 2014* (p. id. 84). The Maui Economic Development Board.
14. Oswal, S., & Koul, S. (2013). Big data analytic and visualization on mobile devices. In *Proceedings of national conference on New Horizons in ITNCNHIT*.
15. Candela, L., Castelli, D., & Pagano, P. (2012). Managing big data through hybrid data infrastructures. *ERCIM News, 89*, 37–38.
16. Menon, A. (2012). Big data@ facebook. In *Proceedings of the 2012 workshop on management of big data systems* (pp. 31–32). ACM.

17. Boellstorff, T. (2015). *Coming of age in second life: An anthropologist explores the virtually human*. Princeton University Press.
18. Silva, T. H., Vaz De Melo, P. O. S., Almeida, J. M., & Loureiro, A. A. F. (2014). Large-scale study of city dynamics and urban social behavior using participatory sensing. *IEEE Wireless Communications, 21*(1), 42–51.
19. Su, T., Zhu, C., Lv, Z., Liu, C., & Li, X. (2016). Multi-dimensional visualization of large-scale marine hydrological environmental data. *Advances in Engineering Software, 95*, 7–15.
20. https://www.journals.elsevier.com/future-generation-computer-systems/call-for-papers/recent-advances-in-big-data-analytics-internet-of-things-and
21. Bashir, M. R., & Gill, A. Q. (2016). Towards an iot big data analytics framework: Smart buildings systems. In *High performance computing and communications; IEEE 14th international conference on smart city; IEEE 2nd international conference on data science and systems (HPCC/SmartCity/DSS), 2016 IEEE 18th international conference on* (pp. 1325–1332). IEEE.
22. https://spark.apache.org/docs/2.0.0-preview/
23. https://erpsolutions.oodles.io/developer-blogs/Event-streaming-KAFKA/
24. Cortes, C., & Vapnik, V. (1995). Support-vector networks. *Machine Learning, 20*(3), 273–297.
25. Song, H., Fink, G., & Jeschke, S. (2017). *Security and privacy in cyber-physical systems: Foundations and applications*. Wiley.
26. Pantelis, K., & Aija, L. (2013). Understanding the value of (big) data. In *Big data, 2013 IEEE international conference on IEEE* (pp. 38–42). IEEE.
27. Chithaluru, P., Al-Turjman, F., Kumar, M., & Stephan, T. (2020). I-AREOR: An energy-balanced clustering protocol for implementing green IoT in smart cities. *Sustainable Cities and Society*, 102254. https://doi.org/10.1016/j.scs.2020.102254
28. Shankar, A., Pandiaraja, P., Sumathi, K., Stephan, T., & Sharma, P. (2020). Privacy preserving E-voting cloud system based on ID based encryption. *Peer-to-Peer Networking and Applications*. https://doi.org/10.1007/s12083-020-00977-4
29. Punitha, S., Al-Turjman, F., & Stephan, T. (2021). An automated breast cancer diagnosis using feature selection and parameter optimization in ANN. *Computers and Electrical Engineering*. https://doi.org/10.1016/j.compeleceng.2020.106958
30. Yadav, S. P., & Yadav, S. (2020). Image fusion using hybrid methods in multimodality medical images. *Medical & Biological Engineering & Computing, 58*, 669–687. https://doi.org/10.1007/s11517-020-02136-6
31. Yadav, S. P., & Yadav, S. (2020). Fusion of medical images in wavelet domain: A hybrid implementation. *Computer Modeling in Engineering and Sciences, 122*(1), 303–321. https://doi.org/10.32604/cmes.2020.08459
32. Dighriri, M., Lee, G. M., & Baker, T. (2018). Big data environment for smart healthcare applications over 5G mobile network. In M. Alani, H. Tawfik, M. Saeed, & O. Anya (Eds.), *Applications of big data analytics*. Springer. https://doi.org/10.1007/978-3-319-76472-6_1
33. Khine, K. L. L., & Nyunt, T. T. S. (2019). Predictive big data analytics using multiple linear regression model. In T. Zin & J. W. Lin (Eds.), *Big data analysis and deep learning applications. ICBDL 2018* (Advances in intelligent systems and computing) (Vol. 744). Springer. https://doi.org/10.1007/978-981-13-0869-7_2
34. Meenakshi, R. A. C., Thippeswamy, M. N., & Bailakare, A. (2019). Role of hadoop in big data handling. In J. Hemanth, X. Fernando, P. Lafata, & Z. Baig (Eds.), *International conference on intelligent data communication technologies and internet of things (ICICI) 2018. ICICI 2018* (Leecture notes on data engineering and communications technologies) (Vol. 26). Springer. https://doi.org/10.1007/978-3-030-03146-6_53
35. Grace Mary Kanaga, E., & Jacob, L. R. (2021). Smart solution for waste management: A coherent framework based on IoT and big data analytics. In J. Peter, S. Fernandes, & A. Alavi (Eds.), *Intelligence in big data technologies—Beyond the hype* (Advances in intelligent systems and computing) (Vol. 1167). Springer. https://doi.org/10.1007/978-981-15-5285-4_9

Untangling E-Voting Platform for Secure and Enhanced Voting Using Blockchain Technology

Muskan Malhotra, Amit Kumar, Suresh Kumar, and Vibhash Yadav

1 Introduction

The blockchain technology was introduced in 2008 by Satoshi Nakamoto [1], who created the first cryptocurrency called Bitcoin. Since technology has been emerging over the years, analysts believe that blockchain will have a wide and impactful scope in the near future. Most likely, people have heard about blockchain, but conceivably, most of them do not consider it, contemplating it a buzzword. But actually, blockchain is a breakthrough technology. The expectation with blockchain technology is so high that it is believed that it may reconstruct most industries in the upcoming years.

1.1 Blockchain Technology

This technology came into existence with bitcoin, a highly popular cryptocurrency. Blockchain can be integrated into several artificial intelligence technologies like neural networks, support vector machines and fuzzy logic. Bitcoin made us take

M. Malhotra · A. Kumar (✉)
Department of Computer Science and Engineering, HMR Institute of Technology and Management, Delhi, India
e-mail: Amit.kumar@hmritm.ac.in

S. Kumar
Department of Computer Science and Engineering, School of Engineering and Technology, Sharda University, Delhi, India
e-mail: sureshkumar@aiactr.ac.in

V. Yadav
Department of Information Technology, Rajkiya Engineering College, Banda, Uttar Pradesh, India
e-mail: vibhashds10@recbanda.ac.in

© The Author(s), under exclusive license to Springer Nature Switzerland AG 2022
F. Al-Turjman et al. (eds.), *Transforming Management with AI, Big-Data, and IoT*, https://doi.org/10.1007/978-3-030-86749-2_3

into account new technology, namely, Blockchain. Blockchain technologies offer to profit applications from sharing economies. When transactions were digitally linked and recorded, distributed ledger came into existence. It is highly shared. The blockchain structure is a type of append-only data structure. In this type of data structure, blocks that are new cannot be altered and eliminated. They can only be written. These blocks are chained in such a way that each block has a hash code. This block is a function of the previous block. This ensures immutability. These blocks ensure the entire history or provenance of an asset. Cryptographically, the chain is signed, duplicated and verified publicly at every transaction [2]. This ensures that none of the individuals can mitigate data that has been written onto a blockchain. A transaction is validated using a consensus protocol and only then added to the blockchain. The consensus protocol ensures that the transaction is the only version of the truth. Each record or transaction is efficiently encrypted.

Hence, Blockchain is nothing but a disseminated database that exists on multiple computers at one time. The technology has been growing since various sets of 'blocks' are added to it. These blocks contain a link to the previous block and a timestamp that forms a chain. The database is managed unanimously by the users as everyone gets a copy of the complete database. All blocks are encrypted, but a special key user will have access to add new records. Hence manipulation in transactions is near to negligible. Blockchain in itself is transparent, secure, and independent.

Blockchain technology is not just a backup mesh, rather it has a lot more to offer. Now, the question arises: What are those key elements that helped blockchain stand out from all other technologies? Why is this technology gaining so much popularity? Let us dive into its key elements to answer these questions.

1.2 Blockchain Working Principle

Blockchain has three pillars: blocks, nodes, and miners.

Blocks: A chain in blockchain consists of blocks which is made up of data, nonce, and hash. There can be multiple blocks in a chain. The data is the building block of any blockchain feature. A nonce is a whole number of 32 bits. The nonce is generated whenever a block is created. It is generated randomly. A 256-bit number wedded to the nonce is known as a hash [3]. Every time a block is generated, a nonce generates the cryptographic hash. Block data is tied to the hash and nonce unless it is mined. The data is also considered signed.

Miners: Mining is a process which helps the miners in creating a chain of new blocks. Every block in a blockchain has its own unique hash and nonce. Every hash also has a reference to the previous block inside the chain. Hence, mining a block is critical, especially when the chain is large. The miner uses special software to solve highly intricate maths problems. These math problems help to generate hash codes that are accepted, by finding nonce. This process is a bit complex as well as complicated because the hash is 256 bits and nonce is only 32 bits. There are approximately four billion possible combinations for hash-nonce that has to be mined before the right one is found. 'Golden Nonce' is the term used, if a miner finds an acceptable combination, and their block is added to the chain.

Making changes in a block requires re-mining of the current block that needs to be changed and all the blocks that come after. This is why it is said that it is notably challenging to manage chains of a blockchain. Finding the golden nonce is a tough job; it requires a tremendous amount of time and effort. A miner is rewarded when a block is mined successfully, and the change is accepted by all of the nodes on the network.

Nodes: Any kind of electronic device that maintains the copies of the blockchain and helps to keep the network functioning is acknowledged as a node. As not a single organization or computer can own the chain, hence it is a distributed ledger via the nodes connected to the chain. Every node has its own copy of the blockchain. The network has the power to permit any newly mined block (algorithmically) for the chain to be verified, updated and trusted. As the blockchain is transparent, hence each and every action is precisely taken into consideration. Each participant that is part of the entire transaction has a unique alphanumeric identification number. Public information is consolidated with balances and checks that create a sense of trust among the user. This maintains the integrity of blockchain.

Blockchain can be considered as the scalability of trust via technology.

1.3   Blockchain Key Elements

- *Distributed Ledger Technology*: Distributed ledger technologies have seen some astonishing technological advancement in the information technology world. Distributed Ledger is a digital system. It is used to record the transaction of assets. It records the transaction at the same time and the details are recorded at multiple positions. Each node processes and verifies each item as it generates a record for each item and creates consensus on each item's veracity. This sort of architecture of computers represents a significant revolution. This significant revolution helps to keep records by changing how information is gathered and communicated.

Unlike traditional databases, these ledgers have no administration functionality or central data storage as it is visible in the centralised model shown in Fig. 1.

All the participants inside a network can access the immutable records of transactions and easily access the distributed ledger. This eliminates the duplicity of efforts present in the conventional business networks; hence, with this shared ledger transactions are recorded only once [4]. It also has the potential to diminish the cost of transactions. The technology makes it challenging to manipulate or attack the system as nodes of the network possess separate records. The information is shared across the network and witnessed, thereby making the system transparent and reliable.

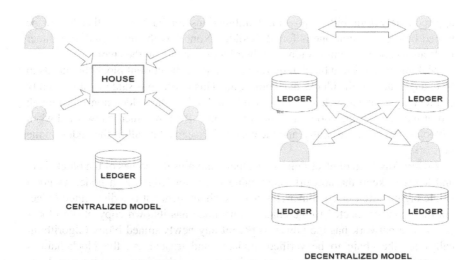

Fig. 1 Centralized vs decentralized ledger

Some aspects of decentralization are:

No malicious activities: Distributed ledger responds drastically to sceptical activity or tampering. Tracking is quite easy through the nodes.

Ownership of verification: This provides fair participation of the user, as on the ledger the nodes act as verifiers [5].

Managership: Every node that is active has to maintain the ledger and participate, so that the features of blockchain are workable.

Quick responses: Removing the intermediate accelerates the system response; thus, any change in the ledger is updated within minutes or even seconds.

- *Records Are Immutable*:

Immutability is the ability of the blockchain ledger to remain stable; this makes the blockchain to remain unaffected and indelible. Using hash values or cryptographic principles, each block of information can be processed efficiently. These blocks of information include facts and transaction details. The hash code value consists of an alphanumeric string generated by each block separately [6]. Each block contains digital signatures and hash values for itself and for the previous block. This makes the block unrelenting and makes them coupled together radioactively. These are the functionalities of blockchain that eliminate any kind of alteration by any intruder.

As we have discussed before, blockchain is distributed and decentralized in nature; therefore, a consensus is made among various nodes that store the replica of data. The originality of data is maintained by these consensuses only. Undoubtedly, immutability is a definite feature of technology. Immutability redefines the data auditing process and makes it cost-effective and efficient, and brings more trust and integrity to the data.

Once the transaction is recorded in the shared ledger, no participant can tamper with the transaction or change the transaction according to his will. Both the transaction records that constitutes an error and the new transaction record that has been added to reverse the error [7]. Even though immutability is beneficial in blockchain, we must remember that this technology has both positive and negative data privacy implications.

- *Smart Contracts*:

Lines of code that are stored on blockchain are known as Smart Contracts. They execute implicitly when deliberate terms and conditions are met. Smart contracts follow simple 'if...then...else' statements.

Smart contracts, also known as the set of rules for blockchain, are automatically executed and stored on the blockchain. These rules help to expedite the transaction. Smart contracts can define conditions for the corporate bond transaction and also include travel insurance terms and much more. Business collaborations are mostly benefited by smart contracts. They are used to enforce different types of agreements so that all participants have an intermediary's involvement [8]. They probably enforce some type of agreement and bring transparency, efficiency, and simplicity to every financial transaction. Smart contracts can be efficiently explained with the help of the supply chain example:

> Buyer B wants to buy something from Seller A, so she puts money in an escrow account. Seller A will use Shipper C to deliver the product to Buyer B. When Buyer B receives the item, the escrow money will be released to Seller A and Shipper C. If Buyer B doesn't receive the shipment by Date Z, the escrow money will be returned. When this transaction is executed, Manufacturer G is notified to create another item sold to increase supply. All this is done automatically.

- *Decentralized Technology*:

By de-centralization, we mean that it does not have an authority to govern and no certain person looks after its frameworks. Nodes group together to maintain the network, making it decentralized. This is the most important characteristic of blockchain technology that helps to work. Blockchain provides access to all the participants as the system is decentralized. Hence, the participants have access to the data which is linked with the web in order to store the assets. Participants can store any kind of information, for example, contracts, cryptocurrencies, important documents or any other valuable digital assets [9]. This is made possible only because the participants have direct control over their data, provided with their private key. Therefore, we can resolve that a decentralized structure gives power to the common people and rights on their assets.

Salient Features of Decentralized Technology
- *Fewer Chances of Failure*: As everything is well-organized in blockchain, it is highly fault-tolerant. Hence, its usual output does not have an accidental failure.
- *Transparency*: The profile of every participant is transparent. Changes are viewable on a blockchain making it more concrete.

- *Genuine by Nature*: This nature of the system makes it distinct and eliminates any kind of actions from hackers to break in.
- *No Third-Party Intervention*: No third-party interference results in no added risk. This nature of the technology removes its reliance on any third party, making it decentralized.
- *Less Prone to Breakdown*: The technology has developed survival techniques for any malicious attack. System attack is expensive and not a solution for hackers. So, it is likely to break down.
- *User Control*: Users have control over their properties. To maintain assets, third-party reliability is removed. All users can maintain and control their assets simultaneously by themselves.
- *No Scams*: There is no chance of scam as the system runs on algorithms. There is a big no for utilizing blockchain to gain personal profit.

Enhanced Security: As the blockchain technology gets relieved of the central authority's need, it does not allow any of its participants to transform any network characteristic for their gain. Another security layer for the system is encryption. This encryption is done with the help of cryptography. Cryptography provides another layer of protection for users, along with decentralization. Cryptography acts as a firewall for attacks having coded in obscure mathematical algorithms. It works on abstraction, that is, concealing the actual information on the chain and hiding the nature of the data [10]. The information undergoes a process that gets the data as input and processes it through various mathematical algorithms. The output that is produced has different values that are always of fixed length. You can think of it as a unique identity that is generated for the data that has been taken as input. Unique hash codes are provided to every block along with the previous block in the ledger. Therefore, tampering with any information on the block will lead to a change in the hash IDs, which is a kind of impossible task. The user is provided with a key which is public to make transactions and a private key to access the data.

- *Irreversible*:

Hashing being complex makes it impossible to alter or modify it. It is even more challenging to get a private key from a public key. If someone wants to forge the network, the hacker is made to modify every aspect of the information stored on every node in the network. All the participants of the node will have a similar copy of the ledger [11]. Accessing such a massive amount of data via hacking is nearly impracticable.

- *Consensus*:

The building block of every blockchain is consensus algorithms. Blockchain technology has efficiently designed architecture, and consensus algorithms are the core of this architecture. Consensus helps to make decisions.

By definition, a consensus makes decision for all the active nodes that are part of the network and makes a group of these nodes. With this feature, nodes instantly agree to the agreement which makes it relatively faster. The importance of consensus is even more when millions of nodes validate a transaction. At this time, a consensus is necessary to maintain the working and flow of transactions and data placidly. You can assume it to be a kind of voting system where the minority has to support the majority that wins.

This feature may lead to trustlessness within the system of nodes. Nodes might not trust other nodes but they can trust the running in core algorithms. Hence, making every decision that is present inside a network is a winning scenario for blockchain. There is a huge variety of consensus algorithms for blockchain around the globe [12]. Every algorithm has a unique way of making decisions along with improving previous mistakes. The architecture creates a realm of fair web. Hence to maintain decentralization, consensus algorithm is must for every blockchain, or else the core value will be lost.

- *Swift Settlement*:

Traditional transaction systems in the bank are quite slow. Processing a transaction takes an immense amount of time even after finalizing all settlements. There are quite a few chances of corruption as well. As compared to traditional baking systems, blockchain offers a faster settlement. With this, the user time is saved as the transfer of money is relatively faster.

1.4 Types of Blockchain

At present there are four types of blockchain networks: private blockchain, public blockchain, consortium blockchain and hybrid blockchain. The right to read or write a blockchain can be restricted or unrestricted to the participants.

1.4.1 Private Blockchains

A private blockchain is a permission-restricted blockchain. Private blockchains are access control-based that restrict the participation of users in the network. It not only limits access to read but the access to write as well. This specifies who can verify their transactions and who have to be provided with the read access only. On a private network, transactions are made more affordable since only a few nodes need to be verified. There is a verification of only trusted nodes that offer guaranteed high-processing power. One cannot be a participant in a private blockchain unless the network administrator invites the user [13]. Validators and participants are restricted to access information in this type of blockchain. There may be one or more administrators in a network, and they rely on third-party transactions. In this type of private blockchain, only the members of the transaction will know about the transaction, whereas any other participant or participants will not know about the transaction. For example, anyone can sell or buy bitcoins without having their identity revealed. Additionally, private blockchains do not allow anonymity, while some of the public blockchains allow. The most common examples of private blockchains are Hyperledger and Ripple (XRP).

1.4.2 Public Blockchains

A public blockchain has open network. Anyone can read, write, or participate in the network and even download the protocol. There is the availability of information in a public domain. The data is accessible to all, and any participant can view, read, and write data on the blockchain, due to its permissionless nature. In a public blockchain, no individual participant has control over the data. Public blockchains are decentralized and immutable [14]. This signifies that an entry cannot be modified or eliminated once it is approved. Another significant factor that distinguishes public blockchain with private blockchain is open reading and writing of data.

The most common examples of public blockchain are Ethereum (ETH) and Bitcoin (BTC). Both the cryptocurrencies can be viewed and used by anyone as they are created with open-source computing codes.

1.4.3 Hybrid Blockchain

A hybrid blockchain consolidates the benefits of both private and public blockchain. On a permissible blockchain, an application or service can be hosted independently for leveraging a public blockchain for security and settlement. To understand the hybrid blockchain model, one must first understand the difference between private and public blockchain. The feature of immutability and trust from a permissionless public network is best for application developers. It still retains the benefit of control and performance, which is provided by a permitted blockchain [15]. It does have use cases in organizations that neither want to deploy proper private blockchain nor do they want to implement a proper public blockchain. They simply want to deploy the best of both the chains. Hybrid blockchain is used in the hyperledger.

1.4.4 Consortium Blockchain

A consortium blockchain is also known as Federated blockchain. It is a creative approach in providing solutions for the demand of the organizations. In this type of technology, some of the aspects of an organization are public and the rest of the aspects remain private. There is no centralized outcome here, as the blockchain is managed by more than one organization.

A consortium blockchain is a partially decentralized blockchain. In this type of blockchain technology, a preselected set of nodes controls the consensus process. Consortium blockchains are often associated with their use in enterprises. Groups of companies collaborate to leverage blockchain technology for improving their business processes. Examples of consortium blockchain include Corda, Quorum, Hyperledger, etc.

1.5 How Secure Is Blockchain?

Blockchain can be secure, trusted, and robust – as long as the technology is appropriately executed. Blockchain technology is transforming the way we do business as it cuts-of-the-middleman – reducing cost, boosting efficiency in numerous vital services. In this way, it has the potential to lead the business efficiently.

But is it secure? Most importantly, can blockchain-based technologies offer privacy and trust to ensure temper-free and private records?

Blockchain is perhaps best understood as a decentralized ledger that reduces costs by removing intermediaries such as adequately decentralizing trust.

A private blockchain controls access to information given to the user making it less transparent than public blockchain. A public blockchain is a transparent ledger, as it is decentralized [16]. On multiple devices an information is stored in encrypted form. Another name for public blockchains is 'censor-free'. It is explicitly resistant to distributed denial-of-service (DDoS) attacks. On the other hand, private networks are more vulnerable to threats as a private blockchain network can be altered by its owner.

Blockchain can upgrade old systems. Voting with the help of blockchain will provide new dimensions to the democratic system. Blockchain will serve the voters as a digital ledger. The technology is known to draw its power from peers, or better known as 'nodes'. These nodes record, process, and verify all transactions across the system. Rather than being stored, a ledger exists on the 'chain' which is simultaneously supported by millions of nodes. Consensus between all auditing nodes on the network will help prevent computers from making undetectable changes to records. Each record is easily verifiable, and the database of transactions is incorruptible, making blockchain encrypted and decentralized [1]. As it does not exist in one place, the network cannot be influenced or taken down by an individual party. As no single authority has access to every feature of blockchain, hence anyone can be part of the network.

The main aim of blockchain is to make the voting process fair and without any third-party intrusion. Blockchain is an easily confirmable and inalterable system. It has capabilities to be an alternative to traditional voting processes. It has solutions that are smart alternatives, and central authority can take these solutions into consideration, in terms of blocks having data in the chain. Blockchain increases the security with which the data is stored in the blocks. It also minimizes the need for an official centre that provides secure elections. Various attempts have been made to tackle the issues of the traditional election. These attempts serve as a benefit to establish an online system and automate the whole process. Estonia became the first country to use electronic voting during its local elections that were held in October 2005. Moreover, Estonia became the first country that legally bounded general elections using the Internet facility to cast the vote. The option of voting over the Internet was available nationally in the local elections. Not only Estonia but also Switzerland and the Austrian Federation of Students in 2009 held their elections electronically. The world on the other hand is accepting this system gradually. Blockchain has impressive attributes to overcome the problems of voter privacy, data integrity and security.

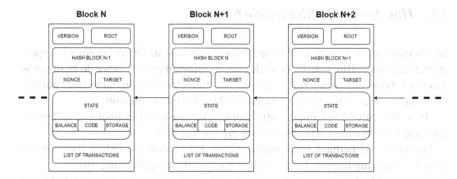

Fig. 2 Block diagram stating the structure of the blockchain

Mobile, as a computing platform, certainly is the future of the democratic voting system. It will provide all new facets to how humans interact with their world. Across the globe, smartphone penetration is expected to exceed 80% or even higher within a few years. Technological challenges and human factors are greater on a mobile device, but the time is propitious for this intersection. 'Mobile voting' can be defined as 'remote voting' and the form factors of 'mobile' will vary widely on different devices around the world [17]. The hurdles will be tough to tackle, and the citizens need to trust their election management bodies.

Sophisticated adversaries provide correlation attacks in the long run. They track the packet sender and receiver. This creates room for hard-to-trace-communication. Receiving messages from multiple senders, shuffling them, and sending them further to the next destination in an arbitrary order is part of such communications. To expedite a decentralized approach in decision-making securely, there is end-to-end encryption that exists within the network protocol of blockchain [18]. With blockchain, we can examine every vote in real time. Blockchain architecture is designed in such a way that it is hack-proof, with the highest security level from the ground up to mass. It renders the hackers' mission virtually impossible making it an ideal platform for voting.

To tackle issues like availability, fairness, anonymity and reliability, various researches have been conducted in order to make the system secure and reliable. Election serves the entire public. Therefore, its design must be in such a way that minimal technical skills and training should be enough. Elections must be flexible enough to assist a large population. Each voter must be given exactly one chance to cast its vote. The system must be competent enough to verify its voters' identity and the certitude of keeping the information secured. To encourage participation, the system must provide an ecosystem for decision-making where authority and resources are shared. E-voting has always remained a cause of discussion both politically as well individually. There is a fervent requirement of strong foundation rules along with mutual understanding among the people (Fig. 2).

2   Literature Survey

In their work, Kanika Garg et al. [19] explained that through the decentralized system, the voting process is made simple, secure and anonymous and hence the focus is constantly drifting. The paper presented a literature review on the techniques used to tackle the challenges of voting. In their article, Hsin-Te Wu et al. [20] have proposed a paper that presents views about the voting system that relies on blockchains to create a trustworthy voting system. To ensure the overall security of a voting procedure, the study also implements a bilinear pairing security mechanism in order to establish both a secret ballot and an open ballot system. According to Ashish Singh et al. [21], the paper focuses on digital e-voting system to solve the security issues and fulfils the system requirements in the blockchain technology. It stated opportunities for a secure e-voting system in any organization or country that needs to be deployed. In 2018, A. Singh et al. [22] explained that the Estonian electronic voting system which is a leading electronic voting system still suffers from universal verifiability issues and may need improvement of its availability. They proposed a blockchain-based electronic voting system in the paper to solve the problems of elections. In their paper, Wenbin Zhang et al. [23] proposed a receipt-free, perfectly verifiable and privacy-preserving peer-voting protocol that can help facilitate voting for peers existing on a blockchain network. Basit Shahzad et al. [24] suggested a framework by using effective hashing techniques to ensure the security of the data. The concept of block creation and block sealing was introduced in the paper. In Nir Kshetri et al. [25], the idea behind the paper was to make eligible voters cast a ballot anonymously using a computer or smartphone. BEV employs tamper-proof personal IDs and encrypted keys, and Friðrik Þ. Hjálmarsson et al. [26] aim to implement a distributed electronic voting system and the application of blockchain as service to implement the electronic voting system. Rumeysa Bulut et al. [27] have proposed a solution to eliminate disadvantages of conventional elections using blockchain.

3   Research Motivation

A voting system contains a set of rules that determine how referendums and elections are conducted and how the results are determined. Government bodies conduct these political electoral systems. These bodies govern every aspect of the election process. When the elections take place, the government bodies take account of different aspects like who can stand as a candidate, who are allowed to vote; every voter has its voter id, how ballots are marked, how the vote is counted, to set a limit on campaign spending and all the other factors are precisely taken into consideration, that can affect the outcome. Constitution and electoral laws define the political electoral system [28]. The election commission typically conducts them. Some electoral systems tend to elect individuals for specific positions such as the prime minister, president, governor, etc. Multiple candidates are also elected for positions such as members of parliament or members of the legislative assembly. Variations are visible at various steps of the electoral system. The most common systems are *first-past-the-post voting*, the *two-round system*, *proportional representation*, and *ranked voting*.

- *First-past-the-post voting*: In this type of electoral system, candidates are selected with the help of votes that are cast by eligible voters. The candidate having the maximum vote in his favour is elected. The voting system can be used for both the divisions, be it single or multi-member elections. The candidate with the highest number of votes is elected. In the multi-member election, each voter casts the same number of votes to fill positions. The candidates who are elected have the highest chances to be placed corresponding to the number of positions.
- *Two-round system*: This type of electoral system is used for electing candidates for the legislative bodies or where presidents are elected directly. In this system, voters cast a single vote for their chosen candidates. The one who gets the majority wins. This is a voting method that is used to elect individual winners. If no individual candidates get a clear majority, then the second round of the voting takes place with either the top two candidates or the candidates who have equal proportions of vote.
- *Proportional Representation*: This type of system's essence is that all votes contribute to result and not just a bare majority. The most prevalent forms of proportional representation require the use of multiple members' voting districts, as no single seat is filled in a proportional manner. Proportional representation categorizes the electoral system in which divisions in an electoral system are reflected proportionally in the elected body [29]. If a certain number of electorates support a particular political party, then a certain number of seats will be won.
- *Ranked Voting*: It is a type of voting system where voters are provided with a ranked ballot to rank their choices in a sequence on the ordinal scale: first, second, third, etc. There are multiple ways in which the ranking can be generated. In the same way, there are multiple ways to count and determine which candidate is (or are) elected. This type of voting system collects more information from the voters as compared to other voting systems. There are different types of ranked voting as well, provided the root process remains the same.

The soundness of election is a matter of national security, in every democracy. Various studies have been working on the possibilities of an electronic voting system, continuously. The goal is to fulfil the needs of the citizens while minimizing the cost of having national elections. With the rise of candidates being elected under democracy, the early voting system was based on a ballot paper system. Replacing the traditional ballot paper scheme with sound election techniques was critical to implement, making the voting process's verification and traceability prone to fraud. Electronic Voting Machines (EVMs) are considered to have flaws. There have been debates about the security and credibility of votes that have been cast through these electronic voting machines. Discussions have been rife on sabotaging the machines, thereby affecting the votes that have been cast on the aforementioned machine.

Satisfying the legal legislator requirements along with establishing an efficient electronic voting system has been a challenge from a long time.

There are numerous aspects of the implementation of blockchain. One such aspect is its implementation in the E-Voting System.

Elections being a huge organization is supposed to provide democracy and democratic rights to the citizens of the country. They play a very crucial role in the life of the citizens and the country. The future of a country lies in the hands of elections. Hence, it is much important for every individual that is part of the election. Even though the election is an organization, they have to be worthy of trust. They must ensure the privacy of votes and security of its voters. Accordingly, the counting of votes under an authorizing body should not be time-consuming. Delay in this counting and declaration of results increase concerns about result manipulation. In order to conduct elections in an efficient manner, we must take into consideration the roles that are involved in the agreement and the different components and transactions that are involved in the agreement process.

These processes include the following:

1. *Planning election in advance*: Structure and planning for elections commence months before the actual voting takes place. The foremost aspect that is taken into consideration is the total population of the district. Having a fair knowledge about the population index and listing down the newly eligible voters increase the expectancy of votes cast. The second factor is the expected turnout. The expected votes cast in the election is known as 'expected turnout'. There have been cases where eligible voters do not cast their votes. The trend can be determined by taking into account past elections. If there was a 25% turnout in the last city elections and no added factors changed the situation, one can figure out that 25% would vote in the elections this time. If due to voters' new eligibility, the turnout increases to be 35%, it further increases the expectancy of the vote cast this time.

2. *Electorate*: The next step is the electorate procedure. All the eligible voters who are allowed to vote is known as the electorate. The election governing body must verify all the eligible voters along with those who will cast their first vote. Providing voter-id cards on time and other aspects like verifying documents of voters, enrolling their names in the voter list and other such details must be precisely taken into consideration. If the factors are neglected, then it may lead to decreasing the expectancy of votes that have been cast. It will also provoke the rights of the voters to vote.

3. *The nomination of candidates*: The nomination of candidates is an important part of the election process. Candidates are nominated by public parties. However, the nominations are regulated by the legislature. To be able to get nominated by the party, the candidate has to provide details to the committee members before the deadline. A majority of selection committee members must support the nomination. The petition, certificate, and nomination application must be filed with the officer specified in the election statute. The nomination officer scrutinizes the papers [30]. If the officer is dissatisfied, he is refused from his candidature. A

candidate can withdraw his nominations even after being granted permission for candidacy. All these factors support the nomination of a candidate.

4. *Scheduling*: Different techniques are used by the parties and the candidates to spread their messages to the voters. Rallies and meetings are organized and processions are carried out. Party leaders, especially the crowd pullers, are assigned to address the public meetings as their task. The candidates do door-to-door canvassing along with the influential personalities in order to attract crowd. Slogans are coined to attract the audiences along with releasing the advertisements to the press before the campaign begins. To highlight the speeches of the leaders and panel discussions of the various party and party members, radio and television are pressed into services. Electronic media plays essential role in creating awareness about the political parties' programs among the people.

5. *Election campaign*: Parties tend to issue their Election Manifestos as a part of their campaign. A manifesto is considered as a statement of great significance. It is a kind of formal statement of the program that consists of a political party's objectives. Reconstructing Centre and State relations, social justice, fiscal reform, economic growth, health, nutrition, education, defence, and world peace are some of the issues that the manifesto deals in. The manifesto contains programs and promises, intending to attract a large number of voters. The party leaders go through a series of interviews to television and newspaper agencies. A wide coverage is given at regular intervals. The most important aspect to note down is that parties are made to stop their election campaign about 48 hours before the time of polling day. Supervision of the whole polling process is done under the guidance of the presiding officer. He ensures that all the electoral norms and practices are adhered.

6. *Declaration of results*: After the polling is done, the voting machines or the ballot boxes are sealed and carried under customs to the counting stations. The counting of the votes begins. It may take some time to announce the results of the elections. After the results are declared, the party that gets the maximum number of votes has to prove its majority. If there are chances that the winning party is unable to prove its majority, the party forms an alliance to prove its majority.

The features of blockchain that we discussed before in the introductory section get operated through advanced cryptography along with providing a level of security which is greater than or equal to any previous known database. Therefore, blockchain technology is considered to be an ideal tool that can be used to modernize the democratic voting process. The aim is to work on solutions in which voters have power, to review the method in which the vote has been cast and that too at any given moment. The method should also have the ability to review the way votes that have been cast for a bill or a particular legislative proposal. This will lead to overall better governance and better outcome of decisions. This will allow people with domain-specific knowledge to present their views liberally. They will have a better understanding of the process, provided the process is transparent, trustworthy, and reliable. Stating some inessentials that thwart the blooming bud of belief for the democratic voting system, these inessentials refrain the process of voting from conducting smoothly.

1. Persuading voters to vote for a particular party
2. Enabling traceability of votes and identifying credentials of voters
3. Inability to ensure trust among the voters that the vote has been counted accurately
4. The third-party intervening and controlling the course of votes being cast
5. Tampering votes and favouritism towards certain beliefs.
6. Single entity control over tally of votes and determining election results
7. Not allowing a certain group of individuals to cast a vote
8. Providing seats to unfit candidates

By overcoming these inessentials, the democratic structure can finally become trustworthy and reliable. Not only these but there have been various aspects of democratic rights that need to be highlighted. With the coming of age technology, the voting system will get a new dimension, thereby overcoming the system's backdrops.

There is another concept called the *Non-Interactive Zero-knowledge proof* [31]. Non-interactive zero-knowledge proof is indirectly related to blockchain. It can be seen as an essential component for satisfying the requirement of the e-voting system. Perhaps, it acts as a building block for conceptualizing blockchain in the electronic voting system. The concept of zero-knowledge proof is a cryptographical method. In this type of method, a party proves to another party that he knows a certain value, without revealing the value. The party that proves is known as 'the prover' and the other party that counters the prover is known as 'the verifier'. A simple example was first demonstrated live by Konstantinos Chalkias and Mike Hearn. Using the example of 'Two balls and the colour-blind friend', the ZKP works as follows: The prover has two balls, one red and one green, and otherwise identical. The verifier (the friend) is colour-blind. To prove that they are differently coloured, you give your friend the balls, who hides them behind his back. Your friend then decides whether to switch the balls between hands or not, and then reveals one of the balls. The prover declares if the balls were switched. By repeating this process, the prover can prove that he can correctly identify the balls, as the verifier confirms that the likelihood of repeated success is halved each time.

A non-interactive zero-knowledge proof is a variant of zero-knowledge proofs. In this type of non-interactive system, the prover and the verifier do not interact with each other. Researchers believe that to achieve computational zero-knowledge without any interaction, a common reference string can be shared between the prover and the verifier. Some studies have also stated that any voter can prove their message's identity and authenticity without a shared public key. This can be achieved with the help of the random oracle model, which in practice can be used as a cryptographic hash function. This scheme is ideally suitable for smart cards, remote control systems, or personal computers, basically in all the microprocessor-based devices [32].

A large aspect of the modern voting system is stuck in the last century. In order to submit paper ballots to local authorities, people have to leave their homes. Any kind of manual evaluation is prone to errors and mistakes. These mistakes may create conditions of distrust among citizens. Moreover, the situation in the current scenario has reached such a level that under conditions of the outbreak, the democratic system faces issues in a pandemic. The recent national elections that were held in South Korea with 44 million voters in the midst of the pandemic define the need for acceptance of e-voting. At times, there have been conditions where it is difficult to put faith in the results due to security gaps. Some of the main issues of the system constitute Trust, Intermediation, Accessibility, and Autonomy. A vote being a small piece of high-value data, systematic infrastructure is extremely valuable and the need of the hour.

A decisive and crucial part of any election is voting. Hence it shows individual rights power along with their concern for the topic. Voting challenges like privacy issues, resistance from fraud, viable and feasible approach, systematic and secure counting of votes must be taken into consideration. A vote is defined as a right to express opinion, choice or wish. It is the right to express one's opinion on how one would like to be governed, in the context of democracy [33]. If this is the primary goal of a vote, is the mechanism we use to capture the user's opinion serving our nation well? So, what is the problem that we need to fix in this? We will be taking into account the different scenarios that the voting process will go through with and without blockchain. This could lead to an affable approach towards the topic. With the help of these scenarios, we can forecast the outcome up to some extent.

1. *The framework of the voting system at present*

The voting system at present goes on too long. Due to which the enthusiasm of voters and elections is drained. The present framework of the voting system is vulnerable to hacking as well. In some parts of the world, electronic voting machines have been doubted to be corrupted. There are some beliefs that revolves around the tampering of voting machine; computer scientists have tampered with the machine to prove that it can actually take place. These facts demolish the faith of the voter on the governing bodies. The other factor that must be taken into consideration is that of the inaccuracy in capturing voters' intent. The touch screen sensors can be knocked out easily just by vibrations or shocks that may occur during machines' transportation. Unless the sensors are re-aligned or corrected at the time of its placement, it may mislead the voter or even misinterpret the voter's intent. For instance, a voter who wants to vote for candidate X, cast a vote for candidate X, but candidate Y would light up instead and then cast for candidate Y. This leads to fraud in the casting process. The machines have always been subject to scrutiny and distrust.

With the help of software programming and coding, any computer software can be generated. The software can easily be corrupted by any programmer who has or knows the source code. It is impossible to test the present voting system for security problems, especially if problems were intentionally introduced and concealed. If the hackers can insert malicious codes to the electronic voting machines' software, it can change the election results completely. They can be triggered by the obscure combinations of keystrokes and commands via the keyboard. If one talks about the physical security of the machines, then there are faults in that too. Many of the DRE (Direct Recording Electronic voting machine) models are under examination regarding the physical hardware controls. It has been surveyed that the EVMs contained loopholes in controls designed to protect the system. All these choices leave people hopeless and disenfranchised about being able to effect change through their votes. In fact, in the long run, these voting systems face much of the backlash. These systems reduce incentives for new candidates to participate, result in fewer parties, increase gerrymandering, give birth to spoiler candidates whose sole aim is to distract the front runners and privatize mass media on political outcomes, the degradation of rights and democracy. None of these are the desired results for a nation or even a political party or candidate's conspiracy. So, what can we do to improve conditions on the voting front? We can change the rules that govern them. We can design the voting process that makes it more expressive and efficient. While there are no perfect systems out there, we can adopt the 'more perfect union' spirit. We need to keep fixing the new problems and keep trying to find better solutions. We have to look ahead and work with new technologies. Walking with the pace of the changing time is the need of the hour. Even when there is much advancement in every field, then why do elections have to be stagnant? We can only make advancements in this by adopting the trends and giving way to new technologies.

2. The framework of the voting system with blockchain

There is a reason why one has to fill out ballots at polling place for our elections. This is because this is our right. The law has given equal opportunity to every eligible voter, to cast his vote. The right to vote is one of the pillars of democracy. So, how to have a better approach towards voting is discussed in this section.

To protect the vote's integrity and the privacy of the voter at the same time can be done with the help of anonymous ballots. Anonymous ballots are the way to go. Digital voting has always been challenging as the verification and validation of each ballot is tough, while keeping them anonymous. These problems of validation of ballots and keeping the voter's privacy into consideration, Blockchain is a step towards the digitalization of the voting process. The privacy issues can be solved with the help of cryptography which is an essential part of blockchain. Blockchain-based voting is already providing new dimensions to elections. At present, US military officials serving overseas are able to cast their votes in their home elections using their smartphones [34]. An amalgamation of blockchain registry and encryption tallies those votes. Countries like Switzerland, Denmark, Brazil and South Korea are already exploring voting techniques with the help of blockchain. Noticeably, Estonia is leading the way ahead, as they have already developed unique ID cards for their citizens to be able to vote. This allows them to cast their vote over blockchain quickly and securely.

There will be a huge and lasting impact on global governance if the essential part of democracy is digitized. Public referendum becomes a feasible option, and citizens can make decisions much more quickly. With the direct democracy by the people, representative democracy may get marginalized. But this is not all, another result is rigging elections; this could become more complex or nearly impossible.

Blockchain voting is similar to analogue voting. The same processes and concepts are applied. The citizen is bound to prove and register their citizenship in a particular jurisdiction to cast a vote. The identity and citizenship can be recorded on the blockchain associated with that user's key. The other most important thing is to cast a vote. This can be achieved in the form of a specially assigned voting token that would be deposited in the user's account. The token will have a time limit after which it gets destroyed via a smart contract or in short, becomes useless. Once the vote is cast, it gets registered on the blockchain where it is verifiable, transparent, and immutable. One can easily declare the results of the election by just counting up the votes [35]. So, now the question arises that if the voting process becomes easy with blockchain, why does voting by blockchain not be implemented everywhere? The reason is there are some complications in this too. One major issue is the verification of the voter's identity. Moreover, we also have to prevent the people who are not a citizens from casting their vote. This is a bit tricky as it depends on the central governing body to verify residency documents, eligibility, and citizenship. Even though this can be achieved with a biometric system's help, it increases the complexity of the model. Once the verification is done, the next step is to separate it from the ballot itself. Most importantly, the key part of democracy is the secret ballots. Nobody should be aware of the fact that to whom the voter has cast its vote. This way they would not be able to influence the vote in any way [36]. The secrecy of ballots can be achieved with the help of zero-knowledge proofs, ring transactions, or various encryption methods. Each method has its technical challenges, benefits, and drawbacks. Proving complete anonymity is still considered the biggest challenge of blockchain voting.

Experts of cybersecurity agree with the fact that blockchain is unhackable. However, the anonymity needed for voting is more difficult to secure, and one has to be very clear that it is not compromised at any cost.

4 Possible Implications

If the blockchain expands in usability as well as popularity for the common people, it has huge implications and is too better than the current voting procedure. It has the power to fundamentally change the way how democracy functions. The blockchain voting provides the benefit of improved transparency. As of now, a voter does not know what happens to the vote once the vote is cast. He has to trust the polling workers that his vote has been counted properly. However, there is no way to judge that the vote has been counted properly. With the help of blockchain, it is possible to track the vote [37]. A history of votes will be generated in the blockchain every time a vote is cast. The side effect of increased transparency is that it reduces fraud. Blockchain has the ability to raise the standard of voting at international platforms, with the communities of the world advocating for blockchain governance in all notions. Blockchain also allows real-time tallying of votes. This indicates that elections can now happen within a shorter time span. In addition to this, if the elections are conducted digitally, then it will lead to less investment in the polling infrastructure. This will completely change the voting procedure for voters. Any voter will be able to cast their vote from anywhere.

Blockchain is not only built specifically for elections but for initiatives within a company which require voting from employees and shareholders. With open vote from shareholders, it may be possible to imagine good decisions at earliest. Increased engagement of the voters will mark the biggest advantage of blockchain-based voting. Easy log in and casting a ballot will be done within minutes, if blockchain makes voting digitally possible from smartphones or computers [38]. It will result in more direct democracy as it would most probably increase the turnout of voters drastically.

Blockchain has tremendous abilities to overcome the problems of data integrity, voter security and privacy. It is impossible to alter any information of a block as it is discerned by other blocks which have the complete set of data.

Blockchain proves to provide an effective and systematic approach that the democratic system requires. A blockchain-based application is not concerned about the security of its Internet connection, because any hacker will not be able to access terminal and hence will not affect other nodes. Independent nodes cryptographically validate every vote, writing it to the ballot box permanently. This makes the system immune to malicious attacks. Counting of votes can be done with absolute certainty, as each ID is attributed to one vote giving zero scopes for tempering the results. Effective submission of votes without revealing the voters' true identity and their political preferences can be considered an auspicious aspect of blockchain. By providing an efficient and irrefutable way to vote from one's phone will encourage participation. Blockchain is paving for a democracy where people will decide the course of policies themselves rather than relying on representatives. A major advancement in rules of the elections will help make such a transparent system. Online voting has its benefits like:

- Ability to vote remotely
- Automatic calculation of results
- Ease in logistical challenges
- Centralized management

Not only elections but also polling, census, and even guided general meetings can be secured with the help of this technology. Blockchain voting software has diverse use cases. Its ability to manage constituencies and engage people is important for the future of society [39]. At present, the technology is in its infancy, but as it matures along with the young voters, it will play the most crucial role in many lives.

Blockchain voting is still not ready or perfect for prime time yet. However, once it gets legitimacy, it is expected to bring an enormous change to the democratic set-up [40]. Making voting more transparent and easier will create a more engaged electorate. Several organizations are currently working and exploring voting on the blockchain. More accessible voting would mean more ongoing referendums on leadership or more frequent representative elections. All these features of blockchain will drastically change the procedure of elections.

5 Future Scope

E-voting using blockchain has a vast scope in the near future. As the technology is constantly advancing, the acceptance of blockchain will soon become much smoother. It will affect the complete outlook of the present scenario of conducting elections. The elections are more transparent, reliable and secure with the use of blockchain technology [41]. Many times elections require the voters to be physically present at the polling booth. This condition results in the reduction of the number of voters who are eligible to vote. It will increase the accessibility of the voters. Blockchain will tackle the convention of reaching out to booths to cast one's vote. Blockchain will help in solve the biggest challenge of Decentralized Voting System as it will pay close attention towards fraud voters. The techniques currently in use in the cryptocurrency systems such as decentralization, anonymity, high security, yet an auditable chain of records, provides wider scope to the use of the blockchain technology in E-voting [42]. Blockchain is not only limited to securing the financial transactions and any type of data transactions as well [43]. The kind of system infrastructure that blockchain will provide is extremely useful for voting. It has been rightly said that "A vote is a small piece of high-valued data," and thus it needs to be supervised with the utmost responsibility.

6 Conclusion

The requirement is to make the entire election process reliable and secure. Voters look up to elections as a medium of expression. The process has to come out clean and valid. The very foundation of an election is shaken even with a small tragic incident, as the voters doubt the creditability of elections. Blockchain will surely be the remedy for the problems prevalent in the present voting systems. The hurdles that make elections a less transparent and secure process will be resolved.

Most importantly, people will have more access to cast their vote, which will further increase the voting process's efficiency. The technology of blockchain is designed in such a way that it provides a refreshing vision to present scenarios. The adaptability of the system, that is, e-voting with blockchain is the biggest concern. Various governance practices of the world will need to come up with solutions to make the blockchain technology more adaptable in terms of voting. "The only thing that remains constant is change" and hence the voting system of the world needs a complete transition of ideas and approach. This will be the required dawn in the world of voting.

References

1. Nakamoto, S. (2008). *Bitcoin: A peer-to-peer electronic cash system.* Working Paper. [Online] Available: https://bitcoin.org/bitcoin.pdf
2. Hafid, A., Hafid, A. S., & Samih, M. (2019). A methodology for a probabilistic security analysis of sharding-based blockchain protocols. In *Proceedings of the international congress on blockchain and applications* (pp. 101–109). Springer.

3. Hafid, A., Hafid, A. S., & Samih, M. (2019). New mathematical model to analyze security of Sharding-based blockchain protocols. *IEEE Access, 7*, 185447–185457.
4. Luu, L., Narayanan, V., Zheng, C., Baweja, K., Gilbert, S., & Saxana, P. (2016). A secure sharding protocol for open blockchains. In *Proceedings of the 2016 ACM SIGSAC conference on computer and communications security* (pp. 17–30). ACM.
5. Kokoris-Kogias, E., Jovanovic, P., Gasser, L., Gailly, N., Syta, E., & Ford, B. (2018). Omniledger: A secure, scale-out, decentralized ledger via sharding. In *Proceedings of the 2018 IEEE symposium on security and privacy (SP)* (pp. 583–598). IEEE.
6. Zamani, M., Movhedi, M., & Raykova, M. (2018). Rapidchain: Scaling blockchain via full sharding. In *Proceedings of the 2018 ACM SIGSAC conference on computer and communications security* (pp. 931–948). ACM.
7. ZILLIQA Team and Others. (2017, 2019). *The ZILLIQA technical whitepaper*. Retrieved September (vol. 16).
8. Harmony Team. *Harmony*. Technical whitepaper. [Online] Available: https://harmony.one/whitepaper.pdf.
9. Manuskin, A., Mirkin, M., & Eyal, I. (2019). *Ostraka: Secure blockchain scaling by node sharding*. arXiv preprint: arXiv:1907.03331.
10. Zochowski, M. (2018). *A highly scalable decentralized transaction system*. Version 1.0. [Online] Available: https://logos.network/whitepaper.pdf
11. Ethereum 2.0. *Ethereum roadmap*. [Online] Available: https://docs.ethhub.io/. Accessed 28 Jan 2020.
12. Buterin, V. *Ethereum sharding FAQ*. [Online] Available: https://github.com/ethereum/wiki/wiki/Sharding-FAQ. Accessed 28 Jan 2020.
13. Madaan, L., Kumar, A., & Bhushan, B. (2020). Working principle, application areas and challenges for blockchain technology. In *2020 IEEE 9th international conference on communication systems and network technologies (CSNT), Gwalior, India* (pp. 254–259). https://doi.org/10.1109/CSNT48778.2020.9115794
14. Dang, H., Dinh, T. T. A., Loghin, D., Chang, E., Lin, Q., & Ooi, B. C. (2019). Towards scaling blockchain systems via sharding. In *Proceedings of the 2019 international conference on Management of Data* (pp. 123–140). ACM.
15. Stegos AG. (2019). *A platform for privacy applications*. White paper version 1.0. [Online] Available: https://stegos.com/docs/whitepaper
16. Al-Bassam, M., Sonnino, A., Bano, S., Hrycyszyn, D., & Danezis, G. (2017). Chainspace: A sharded smart contracts platform. arXiv preprint: arXiv:1708.03778.
17. Wood, G. (2014). *Ethereum: A secure decentralised generalised transaction ledger*. Ethereum project yellow paper (vol. 151, pp. 1–32). [Online] Available: https://gavwood.com/paper.pdf
18. Kim, S., Kwon, Y., & Cho, S. (2018). A survey of scalability solutions on blockchain. In *Proceedings of the 2018 international conference on information and communication technology convergence (ICTC)* (pp. 1204–1207). IEEE.
19. Garg, K., Saraswat, P., Bisht, S., Aggarwal, S. K., Kothuri, S. K., & Gupta, S. (2019). A comparitive analysis on E-voting system using blockchain. In *4th international conference on internet of things: Smart innovation and usages (IoT-SIU), Ghaziabad, India* (pp. 1–4). https://doi.org/10.1109/IoT-SIU.2019.8777471
20. Wu, H., & Yang, C. (2018). A blockchain-based network security mechanism for voting systems. In *1st international cognitive cities conference (IC3)* (pp. 227–230). Okinawa. https://doi.org/10.1109/IC3.2018.00-15
21. Singh, A., & Chatterjee, K. (2018). SecEVS: Secure electronic voting system using blockchain technology. In *International conference on computing, power and communication technologies (GUCON), Greater Noida, Uttar Pradesh, India* (pp. 863–867). https://doi.org/10.1109/GUCON.2018.8675008
22. Salah, K., Rehman, M. H. U., Nizamuddin, N., & Al-Fuqaha, A. (2019). Blockchain for AI: Review and open research challenges. *IEEE Access, 7*, 10127–10149.

23. Zhang, W., et al. (2018). A privacy-preserving voting protocol on blockchain. In *IEEE 11th international conference on cloud computing (CLOUD), San Francisco, CA* (pp. 401–408). https://doi.org/10.1109/CLOUD.2018.00057
24. Shahzad, B., & Crowcroft, J. (2019). Trustworthy electronic voting using adjusted blockchain technology. *IEEE Access, 7*, 24477–24488. https://doi.org/10.1109/ACCESS.2019.2895670
25. Kshetri, N., & Voas, J. (2018). Blockchain-enabled E-voting. *IEEE Software, 35*(4), 95–99. https://doi.org/10.1109/MS.2018.2801546
26. Hjálmarsson, F. Þ., Hreiðarsson, G. K., Hamdaqa, M., & Hjálmtýsson, G. (2018). Blockchain-based E-voting system. In *IEEE 11th international conference on cloud computing (CLOUD), San Francisco, CA* (pp. 983–986). https://doi.org/10.1109/CLOUD.2018.00151
27. Bulut, R., Kantarcı, A., Keskin, S., & Bahtiyar, Ş. (2019). Blockchain-based electronic voting system for elections in Turkey. In *4th international conference on computer science and engineering (UBMK), Samsun, Turkey* (pp. 183–188). https://doi.org/10.1109/UBMK.2019.8907102
28. Yadav, S. P., Mahato, D. P., & Linh, N. T. D. (Eds.). (2020). *Distributed artificial intelligence: A modern approach.* CRC Press.
29. Shen, C., & Pena-Mora, F. (2018). Blockchain for cities—A systematic literature review. *IEEE Access, 6*, 76787–76819.
30. Jaoude, J. A., & Saade, R. G. (2019). Blockchain applications–usage in different domains. *IEEE Access, 7*, 45360–45381.
31. Qiheng, Z., Huawei, H., Zibin, Z., & Jing, B. (2020). Solutions to scalability of blockchain: A survey. *IEEE Access, 8*, 16440–16455.
32. Wang, J., & Wang, H. (2019). Monoxide: Scale out blockchains with asynchronous consensus zones. In *Proceedings of the 16th fUSENIXg symposium on networked systems design and implementation (fNSDIg 19)* (pp. 95–112). USENIX.
33. Rawat, D., Rana, G., Bindra, J., & Kumar, A. (2020). Implementation of blockchain in current transaction systems. *International Journal of Data Structures, 6*(1), 31–59. https://doi.org/10.37628/ijods.v6i1.590
34. Nordrum, A. (2017). Govern by blockchain Dubai wants one platform to rule them all, while Illinois will try anything. *IEEE Spectrum, 54*(10), 54–55.
35. Guangsheng, Y., Xu, W., Kan, Y., Wei, N., Andrew, Z. J., & Ren, L. P. (2020). Survey: Sharding in blockchains. *IEEE Access, 8*, 14155–14181.
36. Wang, G., Shi, Z. J., Nixon, M., & Han, S. (2019). Sok: Sharding on blockchain. In *Proceedings of the 1st ACM conference on advances in financial technologies* (pp. 41–61). ACM.
37. Bansal, P., Panchal, R., Bassi, S., & Kumar, A. (2020). Blockchain for cybersecurity: A comprehensive survey. In *IEEE 9th international conference on communication systems and network technologies (CSNT), Gwalior, India* (pp. 260–265). https://doi.org/10.1109/CSNT48778.2020.9115738
38. Huang, K., Zhang, X., Mu, Y., Rezaeibagha, F., Du, X., & Guizani, N. (2020). Achieving intelligent trust-layer for internet-of-things via SelfRedactable blockchain. *IEEE Transactions on Industrial Informatics, 16*(4), 2677–2686.
39. Shahnaz, A., Qamar, U., & Khalid, A. (2019). Using blockchain for electronic health records. *IEEE Access, 7*, 147782–147795.
40. Mertz, L. (2018). (Block) chain reaction: A blockchain revolution sweeps into health care, offering the possibility for a much-needed data solution. *IEEE Pulse, 9*(3), 4–7.
41. Ferrag, M. A., Derdour, M., Mukherjee, M., Derhab, A., Maglaras, L., & Janicke, H. (2019). Blockchain technologies for the internet of things: Research issues and challenges. *IEEE Internet of Things Journal, 6*(2), 2188–2204.
42. Shankar, A., Pandiaraja, P., Sumathi, K., Stephan, T., & Sharma, P. (2020). Privacy preserving E-voting cloud system based on ID based encryption. *Peer-To-Peer Networking and Applications.* https://doi.org/10.1007/s12083-020-00977-4
43. Yao, H., Mai, T., Wang, J., Ji, Z., Jiang, C., & Qian, Y. (2019). Resource trading in blockchain-based industrial internet of things. *IEEE Transactions on Industrial Informatics, 15*(6), 3602–3609.

Role of Artificial Intelligence in Agriculture: A Comparative Study

Rijwan Khan, Niharika Dhingra, and Neha Bhati

1 Introduction

Agriculture is the basis of the supportability of any economy. It has a key impact on long-term monetary development and basic change, which, however, may differ based on nations. Previously, rural exercises were restricted to food and harvest creation. Yet, over the most recent two decades, it has advanced to preparing, creation, advertising, and conveyance of yields and domesticated animal items. As of now, rural exercises fill in as the essential wellspring of work, improving gross domestic product (GDP), being a wellspring of national exchange, decreasing joblessness, giving crude materials to creation in different businesses, and by and large building up the economy [1, 2]. The Internet of Things (IoT) technology is growing rapidly as a number of physical objects are linked at an exponential pace to the Internet, understanding the definition of the Internet of Things (IoT). The apps provide travel, agriculture, hospitals, factory automation, and emergency response to natural and human-made disasters, where it is impossible to make human decisions. Even cloud computing applications [3] are applied to the agriculture sector in different ways to help it grow. With the worldwide geometric populace rise, it becomes basic that agricultural practices are surveyed with the point of proffering inventive ways to deal with supporting and improving farming exercises. As the development of AI execution in agriculture proceeds, a fascinating and significant inquiry emerges with respect to the jobs of various actors. Given the particular aptitude required to use AI, one may ponder whether it is workable for farmers to build up these abilities in-house. Our investigation recommends that not all farm associations will be required to build up the innovation and calculations. Rather, they might have the option to lease or gain a couple of AI administrations, which would be

R. Khan (✉) · N. Dhingra · N. Bhati
Department of Computer Science and Engineering, ABES Institute of Technology, Affiliated to AKTU Lucknow, Ghaziabad, Uttar Pradesh, India

© The Author(s), under exclusive license to Springer Nature Switzerland AG 2022   73
F. Al-Turjman et al. (eds.), *Transforming Management with AI, Big-Data, and IoT*, https://doi.org/10.1007/978-3-030-86749-2_4

adequate to serve the long haul of destinations of the association. Farmers are finding it difficult to decide the best time to plant seed as climatic conditions change and pollution rises. With the aid of artificial intelligence, farmers can analyze weather conditions using forecasting techniques [4], which allows them to schedule the type of crop that can be grown and when seeds can be sown. New farming companies are rising with creative arrangements, adding to the intensity of the part [5, 6].

1.1 Literature Review

AI could give an edge to the current practices and procedures to accomplish profitability and supportability objectives. For instance, dynamic abilities as AI can help in detecting market value changes of agrarian items and give explicit headings the planting and reaping to stay away from critical crop losses. Early disease identification and altered water system plans could improve general efficiency and viability. Artificial intelligence–empowered weather forecasts give exact, noteworthy bits of knowledge in regard to day-by-day farm exercises continuously.

Programming calculations associate explicit foliage designs with dietary and soil defects, pests, and different diseases. ML is applied in anticipating climate patterns and assessment of farms for pests and weed. One of the most predominant uses of AI is identified with planned and effective irrigation system frameworks. ML calculations dissect the dirt dampness and give sufficient water system procedures relying upon the yield, soil types, and ecological conditions [7–9]. These frameworks, thus, assist in safeguarding with watering and increment yield. The underlying use of AI is by all accounts in developing explicit yields. The outcomes exhibit the utilization of AI strategies to screen the development of yields that are normal and popular and those that require overwhelming and ordinary water system, for example, cotton and grapevines. The goal is to send AI in developing crops that include bigger land mass to empower productive, cost-effective, less-work-concentrated farming practices.

Intelligent AI apps such as the one designed by the authors in [10] are being used by the agriculture industry to help produce healthy crops, manage pests, track soil and developing conditions, organize data for farmers, reduce workload, and enhance a wide variety of agriculture-related activities in the food supply chain. The study of agricultural information from drones and sensors can give valuable data and direction with respect to the water system, crop losses, crop diseases, and pests. In order to reduce energy demand and decouple it from economic development, energy conservation is an essential component of sustainable energy management as implemented in a study by the authors in sensor networks [11]. As a result, increasing agricultural energy production is critical for lowering energy demand and, as a result, prices. Through the exact utilization of manures, pesticides, and systemized water system, AI takes into account the decrease of ecological effects [9, 12].

1.2 How Can AI Bring Revolution in Farming?

The use of AI is changing the method of activities and the executives of homesteads, the key territories of progress being ongoing estimating and reevaluation of business forms. The fast business changes got because mechanical developments drove associations to alter, create, and stretch out their operational abilities to improve effectiveness. That being stated, the essential objective of accomplishing the advantages of AI is intensely dependent on forms used by associations to adequately activate their specialized assets [13]. As per the UN Food and Agricultural Organization (FAO), the worldwide populace will likely stretch around 9.2 billion constantly by 2050. With accessible land assessed at only an extra 4%, it appears that it is not, at this point, a choice to just plant more harvesting fields or breed more steers. What is required is essentially more prominent proficiency than the current cultivating strategies as farmers will be needed to "accomplish more with less."

Here and there the development turns out to be more regrettable by unexpected climate change, pesticide use, and monocropping. Farmers are utilizing AI strategies to defeat these issues and develop food crops by battling against sicknesses and irritations.

2 Applications of Artificial Intelligence in Agriculture

AI is utilized in various enterprises, from assembling to car, one of the fascinating businesses that AI is breaking into is agribusiness; agriculture is a significant industry and an immense piece of the establishment of our economy; as atmospheres are changing and populations are expanding, AI is turning into a technological innovation that is improving and ensuring crop yield [14]. The most well-known uses of AI in the agricultural industry are crop management, weed and pest detection, soil management, etc. Machine learning models are used to follow and anticipate distinctive regular impacts on crop yield, for instance, the atmosphere changes, as shown in Fig. 1.

2.1 Crop Monitoring

Crop monitoring and management begin with planting and proceed with observing development, collecting crops, crop storage, and conveyance of crops. It is summed up as the exercises that improve the growth and yield of agricultural items [6]. Top-to-bottom comprehension of a class of crops as indicated by their planning and flourishing soil type will unquestionably build crop yield. To redesign the yield's proficiency in a way that it supports both farmers and the nation, we have to use the development that assesses the nature of crops and give recommendations. Remote

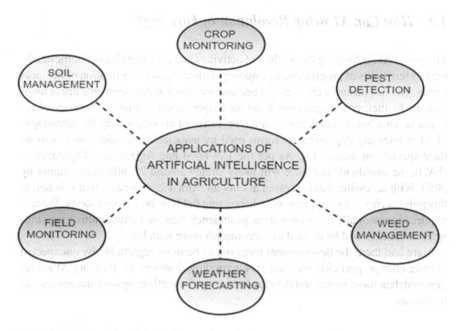

Fig. 1 Applications of artificial intelligence in agriculture

sensors of different sorts are used to assemble the information of yield conditions and environmental changes; then, this information is sent through the framework to the farmers or devices that begin restorative activity. The part of the crop monitoring system is shown in Fig. 2.

2.2 Pest Detection

Pest detection is a significant test in the agribusiness field. The simplest way to control pest infection is the utilization of pesticides. Yet, the extreme utilization of pesticides is injurious to plants, creatures just as people. The procedure of machine vision and advanced image processing [15, 16] are broadly applied to agrarian science, and they have an extraordinary point of view particularly in the plant insurance field, which eventually prompts crop management. Pictures of the leaves influenced by pests are procured by using an advanced camera. The leaves with pest pictures are handled to get a dim-shaded picture, and afterward, highlight extraction and image classification strategies are used to distinguish pests on leaves.

Image processing is examining and controlling graphical pictures from sources, for example, photos and recordings. There are three primary steps in image processing: The first is the transformation of caught pictures into twofold quality that a personal computer (PC) can process; the second is the picture improvement and

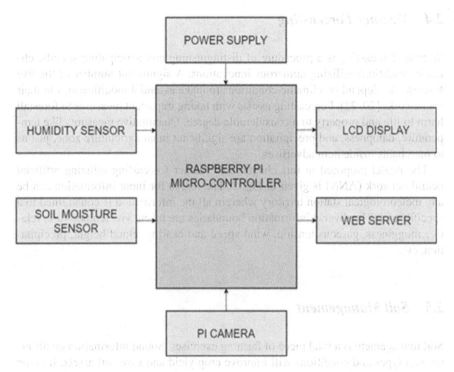

Fig. 2 Block diagram of crop management using the Internet of Things (IoT)

information pressure; and the third is the yield step that comprises of the showcase or printing of the prepared picture [16].

2.3 Weed Management

Weeds are one of the significant obstacles in supporting the harvest profitability. Weeds contend with crops for supplements, soil dampness, sun-oriented radiation, and space and diminish the yield and nature of produce [17–19]. Furthermore, they likewise go about as interchange has for creepy crawly irritations and ailment causing life forms. Weed issues change in various yields, seasons, agrobiological conditions, and the executives' rehearses. Multiple techniques for weed the board in field crops including preventive, social, mechanical, concoction organic and biotechnological, are being utilized with changing level of accomplishment. Because of consistent utilization of a single technique for weed control (particularly herbicide), weeds create opposition and become hard to control.

2.4 Weather Forecasting

Weather forecasting is a procedure of distinguishing and anticipating specific climatic conditions utilizing numerous innovations. A significant number of the live frameworks depend on climatic conditions to make essential modifications in their frameworks [20, 21]. Forecasting assists with taking important measures to forestall harm to life and property to a considerable degree. Quantitative measures like temperature, dampness, and precipitation are significant in an agriculture zone, just as to merchants inside item advertises.

The model proposed in this chapter for weather forecasting utilizing artificial neural network (ANN) is given in Fig. 3. The region for input information can be any meteorological station territory wherein all the information is constrained to a specific area. The diverse information boundaries are taken, viz. temperature, relative mugginess, gaseous tension, wind speed and bearing, cloud height, precipitation, etc.

2.5 Soil Management

Soil management is a vital piece of farming exercises. Sound information on different soil types and conditions will improve crop yield and save soil assets. It is the utilization of tasks, practices, and medicines to enhance soil execution [22, 23]. Urban soils may contain contamination, which can be examined with a conventional soil overview approach. The use of fertilizer and excrement improves soil porosity in total. The use of natural materials is necessary to enhance the quality of soil.

2.6 Field Monitoring

In agricultural field, natural factors, for example, temperature, dampness, sunlight-based radiation, CO_2, and soil dampness, are fundamental components that impact on development rate, efficiency of produce, sugar substance of organic product, sharpness, and so on. If we deal with the previously mentioned natural factors

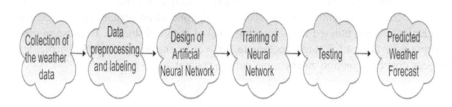

Fig. 3 Weather forecasting using artificial neural network (ANN)

productively, we can accomplish improved outcomes underway of the agricultural items. To check and deal with the development conditions, this chapter recommends the field monitoring server system (FMSS), which can work with sunlight-based force. This FMSS upgraded or improved the force utilization, the portability, and easy-to-understand condition observing techniques [24, 25]. The framework gathers ecological information legitimately acquired from condition sensors, soil sensors, and closed-circuit television (CCTV) camera. To show the area of this framework, a global positioning system (GPS) module is introduced in the framework. At last, we affirmed that the FMSS screens the field conditions by utilizing different offices and effectively works without outside backings.

In this chapter, we have used satellite images for the purpose of field monitoring. The images from the satellite are geo-referenced, and then, they are sent to the communication commission. After this, they pass through the geographic information system (GIS) processing. Then, there is a Web service for online field data collection. It basically includes crop parameters, soil parameters, climatic changes, etc. All the data collected above are being collected in real time. All these real-time data are then fed into a data integration model. Then, there is a Web service that collects the data from the sensors employed in the field. Then, it generates suitable data for a particular field, including crop type, soil type, weather conditions, etc. This method also generates an accurate price for growing a particular crop in the field, including the cost of soil, fertilizers, etc. (Fig. 4)

3 Comparative Study

3.1 Comparison Between Different Crop Management Techniques

- CALEX

- Advantage: It defines adequate scheduling rules for crop management activities.

- Disadvantage: It is time-consuming.
- Artificial neural network (ANN)
- Advantage: It predicts the yield of crop and nutritional disorder in crops.
- Disadvantage: It requires a lot of data for prediction, and it is time-consuming [26].
- Fuzzy logic

- Advantage: It detects the insects that attack the crops.

- Disadvantage: It fails to differentiate between pests and weeds.

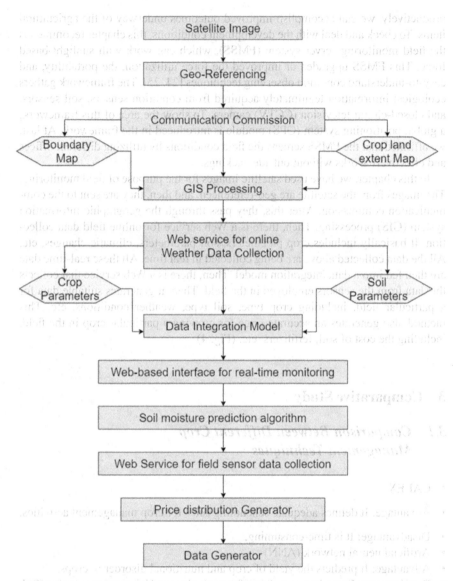

Fig. 4 Flowchart for field monitoring using the Internet of Things (IoT) and Geo-referencing

3.2 Comparison Between Different Soil Management Techniques

- Management-oriented modeling (MOM)
- Advantage: It reduces nitrate leaching, thereby increasing production.

- Disadvantage: It is time-consuming and works only with nitrogen.
- Decision support system (DSS)
- Advantage: It minimizes soil erosion to a large extent.
- Disadvantage: It requires a large amount of data.
- Artificial neural network (ANN)

- Advantage: It is cost-efficient and can predict soil moisture and soil texture.

- Disadvantage: Prediction depends on weather conditions.

3.3 Comparison Between Different Disease Management Techniques

- Computer vision system (CVS)

- Advantage: It supports multitasking and works with a high speed.
- Disadvantage: Its magnitude-based detection affects the quality of some crops.
- Web-based expert system
- Advantage: It is cost-effective and provides a high performance.
- Disadvantage: It depends upon the Internet service.
- Fuzzy logic

- Advantage: It provides more accuracy, and it is eco-friendly.

- Disadvantage: It is expensive and time-consuming.

3.4 Comparison Between Different Weed Management Techniques

- Digital image analysis (DIA)

- Advantage: It provides an accuracy rate of more than 85%.

- Disadvantage: It is time-consuming.
- Support vector machine (SVM)
- Advantage: It detects stress in the crops in a concise time.
- Disadvantage: Only low levels of nitrogen are detected.
- Learning vector quantization (LVQ)

- Advantage: It provides a high weed detection rate.

- Disadvantage: It is quite expensive.

4　Conclusion

Artificial intelligence arrangements need to turn out to be progressively suitable to guarantee that this innovation arrives at the cultivating network. On the off chance that the AI intellectual arrangements are offered on an open-source platform that would make the arrangements progressively moderate, which in the long run will bring about the quicker selection and more prominent knowledge among the farmers. AI presents massive open doors in agricultural applications. Farming arrangements that are AI-fueled empower a farmer to accomplish more with less, upgrading the quality of the crops. In this chapter, the emphasis is on more brilliant, better, and progressively productive yield, developing strategies to fulfill the country's developing food requirements. Worldwide, the population is relied upon arriving at more than eight billion by 2045, which will require an expansion in rural creation by 70% to satisfy the interest. Just about 10% of this expanded creation may originate from unused lands, and the rest ought to be satisfied by current crop production. In this specific circumstance, most recent farming techniques can bring a huge change in agricultural practices. This chapter thought about every one of these viewpoints. It featured the job of different innovations, particularly IoT, to make agriculture more brilliant and progressively effective to meet future desires. For this reason, remote sensors, unmanned aerial vehicles (UAVs), distributed computing, correspondence advancements were discussed altogether.

References

1. Eli-Chukwu, N. C. (2019). Applications of artificial intelligence in agriculture: A review. *Engineering, Technology & Applied Science Research, 9*(4), 4377–4383.
2. Smith, M. J. (2020). Getting value from artificial intelligence in agriculture. *Animal Production Science, 60*(1), 46–54.
3. Shankar, A., Pandiaraja, P., Sumathi, K., Stephan, T., & Sharma, P. (2020). Privacy preserving E-voting cloud system based on ID based encryption. *Peer-to-Peer Networking and Applications*. https://doi.org/10.1007/s12083-020-00977-4
4. Kumar, S., Viral, R., Deep, V., Sharma, P., Kumar, M., Mahmud, M., & Stephan, T. (2021b). Forecasting major impacts of COVID-19 pandemic on country-driven sectors: Challenges, lessons, and future roadmap. *Personal and Ubiquitous Computing*, 1–24.
5. Bestelmeyer, B. T., et al. (2020). Scaling up agricultural research with artificial intelligence. *IT Professional, 22*(3), 33–38.
6. Balaji, G. N., et al. (2018). Iot based smart crop monitoring in farm land. *Imperial Journal of Interdisciplinary Research (IJIR), 4*, 88–92.
7. Yadav, S. P., Mahato, D. P., & Linh, N. T. D. (2020). *Distributed artificial intelligence: A modern approach* (1st ed.). CRC Press. https://doi.org/10.1201/9781003038467
8. Liakos, K. G., et al. (2018). Machine learning in agriculture: A review. *Sensors, 18*(8), 2674.
9. Ampatzidis, Y. (2018). [AE529] applications of artificial intelligence for precision agriculture. *EDIS, 6*, 1–5.
10. Kaiser, M. S., Mahmud, M., Noor, M. B. T., Zenia, N. Z., Mamun, S. A., Mahmud, K. M. A., Azad, S., Aradhya, V. N. M., Punitha, S., Stephan, T., Kannan, R., Hanif, M., Sharmeen, T., Chen, T., & Hussain, A. (2021). iWorkSafe: Towards healthy workplaces during COVID-19

with an intelligent pHealth app for industrial settings. *IEEE Access*. [Online] pp. 1–1. Available at: https://ieeexplore.ieee.org/document/9317697. Accessed 24 Jan 2021.

11. Stephan, T., Al-Turjman, F., Joseph, K. S., Balusamy, B., & Srivastava, S. (2020). Artificial intelligence inspired energy and spectrum aware cluster based routing protocol for cognitive radio sensor networks. *Journal of Parallel and Distributed Computing*. https://doi.org/10.1016/j.jpdc.2020.04.007

12. Bu, F., & Wang, X. (2019). A smart agriculture IoT system based on deep reinforcement learning. *Future Generation Computer Systems, 99*, 500–507.

13. Sowmya, K., & S. Anuradha. (2020). Era of artificial intelligence-prospects for Indian agriculture. *Think India Journal 22.44*, 13–20.

14. Sheikh, J. A., Cheema, S. M., Ali, M., Amjad, Z., Tariq, J. Z., & Naz, A. (2020). IoT and AI in precision agriculture: Designing smart system to support illiterate farmers. *Advances in Intelligent Systems and Computing*, 490–496. https://doi.org/10.1007/978-3-030-51328-3_67

15. Yadav, S. P., Agrawal, K. K., Bhati, B. S., et al. (2020). Blockchain-based cryptocurrency regulation: An overview. *Computational Economics*. https://doi.org/10.1007/s10614-020-10050-0

16. Heeb, L., Jenner, E., & Cock, M. J. W. (2019). Climate-smart pest management: Building resilience of farms and landscapes to changing pest threats. *Journal of Pest Science, 92*(3), 951–969.

17. Liu, B., & Bruch, R. (2020). Weed detection for selective spraying: A review. *Current Robotics Reports, 1*(1), 19–26.

18. Khan, R., et al. (2020). Social media analysis with AI: Sentiment analysis techniques for the analysis of twitter covid-19 data. *Journal of Critical Reviews, 7*(9), 2761–2774.

19. Westwood, J. H., et al. (2018). Weed management in 2050: Perspectives on the future of weed science. *Weed Science, 66*(3), 275–285.

20. Chattopadhyay, N. (2017). Combating effect of climate change and climatic variability on Indian agriculture through smart weather forecasting and ICT application. *Agriculture Under Climate Change: Threats, Strategies and Policies*, 3–8.

21. Khan, R., Amjad, M., & Srivastava, A. K. (2017). Generation of automatic test cases with mutation analysis and hybrid genetic algorithm. In *3rd international conference on computational intelligence & communication technology (CICT)*. IEEE.

22. Yadav, S. P., & Yadav, S. (2020). Image fusion using hybrid methods in multimodality medical images. *Medical & Biological Engineering & Computing, 58*, 669–687. https://doi.org/10.1007/s11517-020-02136-6

23. Fernandez, G., et al. (2020). Smart soil monitoring and water conservation using irrigation on technology. *Indonesian Journal of Electrical Engineering and Computer Science, 19*(1), 99–107.

24. Paccioretti, P., Córdoba, M., & Balzarini, M. (2020). FastMapping: Software to create field maps and identify management zones in precision agriculture. *Computers and Electronics in Agriculture, 175*, 105556.

25. Jihua, M., et al. (2018). A remote sensing-based field monitoring system to support precision agriculture. *Chinese High Technology Letters, 6*, 2.

26. Aggarwal, A., & Kumar, M. (2020). Image surface texture analysis and classification using deep learning. *Multimedia Tools and Applications*. https://doi.org/10.1007/s11042-020-09520-2

Big Data: Related Technologies and Applications

Geetika Munjal and Manoj Kumar

1 Introduction

Data is described as "fundamental values or facts" taken from any person or agencies. Big Data is a type of data with an enormous volume. Big Data is the terminology used to express the immense volume and rapidly increasing data collection over time. Such data is so vast and complicated that no traditional data administration tool can save or process it efficiently. Big Data is hard to track. It represents the sum of all digital information, which has the inconvenience of storing, transporting and analysing. It is extensive and the technology is so tremendous that we have the challenge of creating today and the next generation's data storage tools and technologies.

In the epoch of Big Data [30], by analysing large quantities regarding data availability, it is possible to make rapid progress in many experimental methods and improve many organizations' efficiency and achievement. This creates unique possibilities for Big Data firms to gain more profound, sharper perspicacity that will empower decision-making, enhance the client's expertise and stimulate discovery. Firms are so surprised by the volume and type of data plus their activities that they strive to save data – evaluate, understand, and represent that in a significant way. The word "Big Data" is more than organized and agreement-aligned data. This involves videos, RFID records, communal schmoozing communications, demodulator channels, hunt indexes, natural happenings, pharmaceutical examines, "data

G. Munjal (✉)
Department of Computer Science, Amity University, Noida, Uttar Pradesh, India
e-mail: munjal.geetika@gmail.com; gmunjal@amity.edu

M. Kumar
School of Computer Science, University of Petroleum and Energy Studies,
Dehradun, Uttarakhand, India
e-mail: wss.manojkumar@gmail.com

exhausts" – web surfers that track clicks through the Internet. Big Data technologies complement Business Intelligence (BI) instruments to open content from company's knowledge. It typically executes organized analytics and behind-the-scenes reflector of business administration, whereas Big Data analytics gives a forward-looking aspect, permitting corporations to forecast and perform on future events [1]. Big Data is a relative expression that illustrates a situation where the volume, velocity, and variety of data exceed the company's depository or its ability to make precise and appropriate decisions. Big Data, similar to Business Intelligence, Business Analytics, and Data Mining, has remodelled BI from reporting and determination support to forecast and next-move decision-making [2, 3].

Businesses use Big Data to chase gains, and authorities use it to serve the public welfare. Big Data provides the tools, techniques, technologies; IT structures to increase the exponential volume of diverse information and improve organizations' innovation and competitiveness in implementing sound and timely management decisions. Based on previous literature, it appears that factor models are the most common and widespread methods currently used for Big Data forecasting techniques such as the one seen in [28], whereas neural networks and Bayesian models are two other popular options.

The application of Big Data [31] can significantly benefit a small and medium-sized company, with businesses committed to the resources to execute Big Data technology. To make most of the Big Data, organizations need to develop their IT infrastructure to manage these new huge volumes, high accelerations, and various data origins and combine them with preceding enterprise data to analyse. Miscellaneous queries can be promptly resolved by applying Big Data and worldly wise analytics in a classified, memory, and lateral environment. The drift toward visualization-based data exploration tools can be realized by any business that wants to get the most value out of Big Data. Urban Big Data includes various types of datasets, such as air quality data, meteorological data, and weather forecast data. Raheja et al. [29] modelled the simulation of real-time air quality of any arbitrary location given environmental data and historical air quality data from very sparse monitoring locations.

Big Data is categorized into three elements: (a) data is diverse, (b) data cannot be classified as general relational databases, and (c) data is produced, compiled, and processed swiftly. Big Data is optimistic for business purposes and is immediately growing as a part of the IT industry. This has created considerable interest in various sectors such as healthcare mechanisms making, banking activities, social media, and satellite imaging. Traditionally, data was stored in an extremely organized arrangement to increase its information content. However, modern data quantities are driven by structured and semi-constructed data. Thus, the interchange between structured and unstructured data for analytics in relational systems of database management interrupts end-to-end processing.

The data collected at tremendous growth rate produces several key issues and challenges described, namely, faster data development, transfer velocity, assorted data, and security concerns. However, the advancement of data storage and tunnelling technologies enables the protection of this enhanced data. In aforementioned conservation process, the characteristics of the data produced by the companies are altered. However, Big Data is yet at an early platform and has not been generally analysed.

2 Literature Review

In the information age, huge amounts of data are available for decision-makers. Big Data relates to datasets that are not only large but also big in size and variability, making them tricky to manage using conventional instruments and methods. With the speedy increase of data, there is a necessity to study and provide solutions to manage and gather value and information from these datasets.

This chapter surveys the most modern methods ripened for Big Data. Its mission is to assist you in choosing the correct consolidation of various technologies of Big Data and adapt them to the needs of their technology requirements and particular applications. Besides providing a global aspect of the major Big Data technologies, it also compares them to different system layers, such as Data Storage Layer, Data Processing Layer, Data Query Layer, Data Access Layer, and Management Layer.

This research's main intention is to examine the multiple Big Data technologies that can be applied to manage enormous volumes of data from various roots and improve the overall operation of the systems and its applications.

Big Data Technologies
Various Big Data technologies include:

(i) Column-based database: Traditional, queue-based databases are great for online trade processing at high modernized speeds, but as data volumes increase, they decrease in query performance, and data becomes more unorganized. Column-based databases stock data with a target on columns rather than rows, which allows for massive data squeezing and much faster inquiry times. The drawback of these databases is that they only permit batch updates with a much more delayed update time than traditional standards.

(ii) Schema-less database or NoSQL database: Several database types suit into this category, namely, the Value Store and Document Store, which focus on the depository and recovery of massive volumes of partly organized, or organized data. In NoSQL, this refers only to SQL, which covers a range of contrasting database technologies. The NoSQL database fields for processing relational ancestors, active, semi-structured data with minimal dormancy make them well suited to the Big Data ecosystem. NoSQL is generally described as operational and analytical. NoSQL is a custom function criterion for enhanced auxiliary functions based on an incomplete standard, where data can be processed at unrealistic times. Other big names in the NoSQL field are Cassandra, Oracle NoSQL and MongoDB.

(iii) Mass Parallel Processing (MPP) technologies process large volumes of data in parallel. A number of processors, each with their control system and memory, work in contrasting parts of the identical program. MPP is a static process that requires a static database function between all the processors involved. During MPP, messages are exchanged between processors through the interconnection of data paths. For applications that allow multiple databases to be searched in parallel, an MPP system is considered superior to a symmetrically parallel system (SMP). There are examples of MPP architecture like supercomputers.

(iv) High-Performance Computing Cluster (HPCC) is a free source platform utilized for computing and rendering services to deal with Big Data workflows. The HPCC data design is determined by the customer end in line with the specifications. The HPCC system is projected and then planned to handle the most perplexing and data-intensive analytics-associated difficulties. The HPCC system is the only platform for a particular programming language for single architecture and data simulation. The HPCC system is planned to analyse large volumes of data to solve the perplexing problem of Big Data. The HPCC system is based on an organization control language with the Declarative and Procedural Nature of programming language.

(v) Hadoop is a free source software structure that processes massive amounts of data and processes large amounts of data. Hadoop presents the tools needed to develop and run applications. The data is divided into blocks and stored on multiple connected nodes that work together; this set-up is suggested as cluster.

The Hadoop cluster can traverse thousands of nodes. Calculations run across the cluster in parallel, which indicates that the task is split between nodes in the cluster. The Hadoop structure is penned in Java, which permits custom-written programs to locate system composed programs or a different language to refine data in order across millions of commodity servers. Hadoop employs a set of nodes to run the MapReduce programs in parallel. MapReduce program comprises two stages: the first map phase contains the processed input data, and the second reduction stage integrates the mediators in the result. To run MapReduce programs, each cluster node has a local CPU and a local file system [4, 9]. The data is divided into databases, collected in local files of different nodes, and evaluated for reliability. Local files create a file system termed as Hadoop Distributed File System (HDFS). Each cluster has several nodes ranging from thousands of machines to thousands of nodes. Hadoop can be combined with a fixed set of failover scenarios. Hadoop Ecosystem is a platform or suite that renders multiple assistance to resolve Big Data queries. This includes Apache designs and different commercial instruments and clarifications.

2.1 Components of Hadoop

The four main components of Hadoop are HDFS, MapReduce, Yarn and Hadoop Common in Fig. 1. Many tools or clarifications can be used to strengthen or promote these key components.

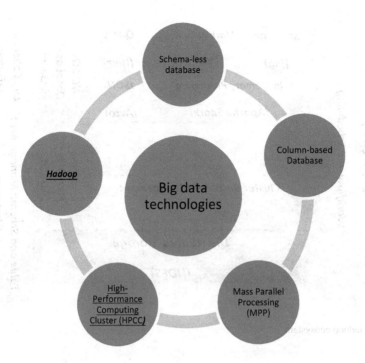

Fig. 1 Big data technologies

2.1.1 Management Layer

It helps in coordination and workflow, Zookeeper, Avro and Oozie as shown in Fig. 2. Zookeeper is a classified, public reference integration service for shared applications. It comprises of master and slave nodes and store composition data. Zookeeper promotes high efficiency and accessibility of data. It clarifies distribution programming and ensures a secure distribution area. The Zookeeper server runs in multiple groups. By its manageable port, Zookeeper facilitates quick, extensible, and secure high-performance computing framework coordination services for distributed systems [5]. For example, it offers a distributed set-up configuration management service, a naming service to find machines in massive clusters, a replication sync service to protect against loss of data and nodes, and a serialized access to resource-sharing locking service.

Avro is a time-worn method call and data serialization structure formed inside Apache's Hadoop design. From defining data types to organizing the data in a compressed binary composition, all activities are usually done by Avro. Data of Avro is saved in a register, with its plot collected so that any program can process it. Oozie: Organizing Apache Hadoop jobs is accomplished by Apache Oozie, which is a Java web application. Oozie couples diversified functions into one relevant task in a row. Oozie can schedule specific jobs for systems such as Java programs or shell scripts. It is an extensible, trustworthy, and extendable method [6].

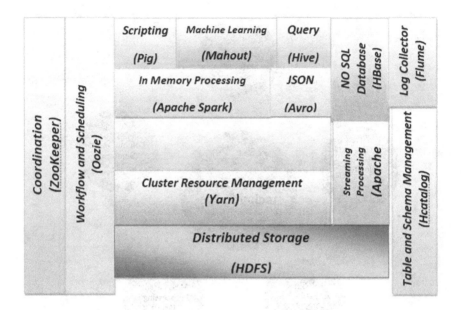

Fig. 2 Hadoop ecosystem

2.1.2 Data Storage Layer

HDFS (Hadoop Distributed File System) and HBase, where HBase is a dispersed, capable, scalable, NoSQL database that sits on the head of HFDS. It stores organized data in tables that contain lots of columns and rows. It can be used on Power Historical Discovery through huge datasets, exceptionally when the aspired data is in large amounts or inconsistent data. HBase is not a unique database and was not constructed for advertisement deals and additional real-time applications. It is available by the Java API and includes the ODBC and JDBC drivers [7]. When it comes to collecting massive volumes of data, we need more than one system, the main network system. Data can then be separated from multiple machines connected to each device over the network. This type of management is called a distributed file system to stock large amounts of data. Hadoop has its classified file system, called HDFS. This is the core of the Hadoop framework. It also eliminates excessive bandwidth across the cluster. It is a Java-based distributed file system that can stock all varieties of data without a previous company. There are high chances for the file system to be corrupted due to software bugs or human errors during software upgrades. The main purpose of creating snapshots in HDFS is to reduce the loss for data stored in the system during the upgrade.

2.1.3 Data Querying Layer

Pig, JAQL, and Hive; Apache Pig enables Apache Hadoop customers to draft complicated MapReduce changes practicing simply the expression of scripting termed as Pig Latin. Pig is an expensive program where the MapReduce structure is built, applied with the Hadoop platform. As in HLL by Pig, data accounts are examined, which is a top profile data processing method. The characteristic of the Pig is flaccidity, simply programmed, and self-optimizing. JAQL, the language announced on Hadoop, provides an inquiry language and promotes large-scale data processing. This transforms high-level inquiries into MapReduce tasks. It is intended to question semi-organized data according to the format of JSON. Such characteristics assure the processing of data, storage, translation, and transformation of data in JSON format. Facebook originally evolved hive. Organized data is processed by hive, a data depository foundation device. Hive is responsible to sum up Big Data and makes it simple to query and analyse. Hive generates its inquiry language called hiveQL. Hives are speedy, extensible, and compatible [8, 9].

2.1.4 Data Access Layer

It includes Data Ingestion; Chukwa is a framework for collecting data and analysing deal with MapReduce and HDFS. This frame is currently advancing beyond its developmental stage. Chukwa receives and processes data from distribution systems and reserves them in Hadoop. As an autonomous module, Chukwa is involved in the distribution of Apache Hadoop. It is set up on the upmost layer of the HDFS and MapReduce structure and acquires Hadoop's extensibility and hardihood. Flume is a disseminated, secure and accessible object to efficiently collect, integrate and move shells of log data. It has a convenient and adjustable architecture based on gushing data flow. This is a powerful dysfunction with tolerance and recovery mechanism with a tunable reliability mechanism [10]. It uses an asymptomatic, extensible data form that enables online application. When the run is off the line, it produces a run load file log. Whether this happens hundredth or millenary of times a day, huge amounts of log files can transmit data. The Flume tool can be stored for months or years of product runs for Apache Storm or similar day analysis in HDFS.

2.1.5 Data Streaming

It includes storm and spark where storm is a free source distribution system that holds the benefit of dealing with real-time data processing. The storm depends on the topology, which contains a network of spout, bolt, and streams. The bolt is utilized to process the input stream to create the output stream. Therefore, the storm is suitable for conversions on streams using "spouts" and "bolts".

The storm is simple to use, fast, extensible, and fault-tolerant system, and when more than one process fails, the storm instinctively relaunches. If the process crashes frequently, it would be sent back to different machines and would be restarted again by storm. Real-time analysis, online machine learning, constant computing, and dispersed RPC are several such situations in which it can be practiced. To increase its performance, Spark is established on an in-memory system and is similar to Hadoop. It is a verified analytics platform that assures quick, simple to practice, and adaptable computing. Spark makes a complicated analysis of vast data collections through the In-Memory MapReduce system. The Spark Project includes task scheduling, memory administration, error retrieval, cooperation with storehouse systems, and more.

2.1.6 Data Processing Layer

It contains Hadoop MapReduce which provides the software infrastructure to make writing applications easier. This fault-tolerant trusted processor processes massive volumes of passive data in correspondence to a huge number of Ticklester Commodity hardware. The MapReduce job typically splits the input data into autonomous components, which are treated in parallel through map functions. Both input and output functions are saved in the file system. Scheduling tasks outline, supervise and perform indelible tasks [11, 12, 14]. The series MapReduce indicates less work is constantly done following the map job. The MapReduce framework's upper hand is cost-effectiveness, flexibility, and scalability due to its underlying parallel processing architecture.

2.1.7 Hadoop

YARN is an iconic part of the public source Hadoop platform for Big Data analytics accredited by the nonprofit Software Foundation. The essential components of Hadoop comprise the Central Library System, the Hadoop HDFS File Handling System, and Hadoop, MapReduce, which uses data to contain resources. Hadoop Yarn is defined as a grouping platform that manages assistant resources and scheduling assignments.

2.1.8 Mahout

Apache Mahout is a free source project that is mainly applied in the construction of extensible machine learning algorithms. It executes successful machine learning methods: advice, categorization, assembling. It is split into four central groups: grouping, refining, classification, and drilling of parallel periodic models. The Mahout Library refers to a sub-community that can be run in distribution mode and accomplished by MapReduce [13].

3 Applications of Big Data

Big Data is almost ubiquitous. Every business can implement Big Data analytics, such as health or normal living standards. Big Data is a track that can be applied in any field, and this huge amount of data can be used for one's benefit. The main applications of Big Data are posted beneath [15–17].

In *Agriculture*, various biotechnology companies uses sensor data [25, 28] to increase obtaining efficiencies. It collects plant tests and re-examines how plants respond to different changes in conditions. Its information is gathered around it for the quality and temperature of various plants, water level, soil system, growth, yield and quality of each plant in a proven ground. These recreations are approved to ensure the ideal environment for the correct quality sorts.

In *Finance,* relevant companies use external praise score when evaluating new acclaimed applications. Besides, banks are currently using their commendation score checks, which use a wide range of information for available customers and checks information of balances, charge cards, home loans, and corporations. Any monetary institution's performance depends on its data, and securing that data is one of the most challenging hurdles faced by any monetary institution. Data is the secondary commodity to them after wealth. Even ere Big Data became popular, the finance industry had previously conquered the tech sector. Additionally, financial institutions are amongst the lighthouse customers of Big Data and analytics. A big investment is at the heart of two of the most popular jargons in digital banking and payments. Big Data improves monetary institutions' core domains, such as deception discovery, uncertainty interpretation, algorithmic trading, and client fulfilment [18].

In *Banking and Securities Industry*, using customer data can also lead to privacy issues [27]. By looking at the incompatible parts between it, the exact bits of information, Big Data analytics uncover sensitive personal data. This shows that many financiers are wary of using Big Data because of isolation issues. Besides, the outsourcing of information research implementation or the sharing of customer communication and the product of a happy understanding opens up workplace security threats [26]. The Securities Exchange Commission (SEC) utilizes Big Data to observe the monetary market action. They are presently utilizing network analytics and universal language processors to capture unlawful commerce action in commercial markets. The industry depends on Big Data for exposure assessment: money lending, enterprise exposure administration, "know your customer" and demand reducing deceit. Big Data suppliers practicing exclusively in this industry involve Panopticon Software, Streambase Systems, Nice Optimize, and Quartet FS.

In *Education*, Big Data is the key to shape people's future and can remodel the education system for the betterment of the world. It is not just re-awakening collegiate skills, but also non-collegiate skills such as interpersonal abilities. Some best educational institutions are practicing Big Data as a means to reinvent their scholastic curriculum. Educational institutions can also trace student dropout rates and take the necessary steps to diminish this rate as much as feasible. In the state of the differential application of Big Data in education, it can also be utilized to estimate professor potency to assure a delightful experience for learners and tutors. Professor performance can be accurately measured and marked upon numbers obtained by the student, the number of students, student goals, behaviour analysis, and countless additional factors. At the state level, the USA has been working to develop analytics to help students use the online curriculum to get the right curriculum. Among the Big Data suppliers in the industry are Newton and My Fit/Navion.

Manufacturing and natural resources: In the natural supplies industry, Big Data provides predictive modelling to be practiced to consolidate data and uniting massive amounts of geospatial data, graphical data, document, and transient data. Big Data has been adopted to address present production hurdles and achieve competitive benefits along with other advantages. Predictive manufacturing provides useless time near to zero and transparency. Huge amounts of data and sophisticated assessment tools are needed to be useful for the systematic process of data. The main advantages of practicing Big Data applications in the production industry are: i. Goods condition and bugs tracking ii. Stock preparation .iii. Production process and fault tracking iv. Yield anticipating v. Raising power productivity vi. Analysis and simulation of the latest production methods [21].

Government: In public services, Big Data has a broad variety of uses, including power research, economic market analysis, duplicity disclosure, health-associated analysis, and environmental security. Big Data is being practiced by the Social Security Administration (SSA) in the study of huge numbers of social disadvantage rights in the manner of structured data. Analytics can be utilized to process pharmaceutical knowledge faster and more accurately to make more agile decisions and identify unusual or counterfeit cases [22].

To recognize and analyse food-associated sicknesses and disease patterns, the Food and Drug Administration (FDA) uses Big Data. This enables it to respond quicker, leading to breakneck healing and less death of people around the world. Big Data is practiced for multiple cases by the Department of Homeland Security. Big Data from numerous regime companies are analysed and utilized to shield the homeland. Among the Big Data suppliers in the industry are Digital Reasoning, Socrates, and Hewlett-Packard.

Transportation industry, governments practice Big Data [19]: gridlock control, path outlining, smart transportation methods, bottleneck administration. Private-sector usage of Big Data in transportation: revenue administration, high-tech improvements, coordination, and competing for advantage. Personal usage of Big Data involves time-saving for energy savings and travel arrangements in tourism. Qualcomm and Manhattan Associates are among the industry's Big Data providers.

Energy and utility industry: Intelligent meter readers permit older meter readers to collect data every 15 minutes rather than once a day. This comminuted data is applied to better examine utilities, allowing more loyal user response and greater command over performance. In monopoly companies, Big Data application also enables better resource and labour pool administration, which can be used to identify shortcomings and correct them as quickly as feasible ere a complete breakdown occurs. Alstom Siemens ABB and Cloudera are among Big Data suppliers in the industry.

Big Data Analytics has advanced healthcare by rendering personalized antibiotic and prescription analytics. Researchers are mining data to see which treatments are more beneficial for singular conditions, recognize drug aftereffects and other relevant knowledge that can assist patients and trim expenses. The amount of data along with mHealth, eHealth, and wearable technologies is growing at an accelerated increase rate. This comprises computerized health report data; patient produced data, detector data, and different types of data. By mapping healthcare data with geographic datasets, it is feasible to prognosticate disease progression in particular regions. According to estimates, it is easy to diagnose and prepare serums and vaccines. Some hospitals, such as Beth Israel, utilize data gathered from millions of patients from a cell phone app, to permit physicians to oppose multiple medical/ laboratory tests for hospitalized patients to oppose evidence-based medicine. The University of Florida has used open public health data and Google Maps to build optical data that can help you quickly and effectively analyse healthcare information care used to detect the spread of chronic diseases. Hummedica, Explorer, and Cerner are among major data suppliers in the industry.

Media and entertainment: Several organizations in the media and entertainment industry are encountering innovative business models – creating, marketing, and distributing their content. This is due to the innovation of current users and the need to access content on any gadget, anytime, anywhere. Big Data presents facts regarding diverse people. Presently, publishing conditions are modifying ads and content to attract customers. This information is collected through several data tunnelling operations [22]. Media and entertainment industry is being benefited from Big Data utilization by foretelling what the audience wants, schedule optimization, growth of acquisition and retention and goal of advertising.

Internet of things (IoT) and Big Data: IoT is one of the major businesses for Big Data applications. Due to the immense diversity of objects, applications of IoT [29] are constantly unfolding. Nowadays, several Big Data applications are being supported by logistics companies. It is likely to trace the location of vehicles with sensors, wireless adapters, and GPS. Data collected from the IoT device provides a mapping of device interconnectivity. Several organizations and authorities have used it to improve the performance with the help of such mapping. IoT is also widely used as a tool for audiovisual data collection, and this audiovisual data is used in pharmaceutical and production contexts. Therefore, such data-driven applications permit organizations to monitor and supervise employees in addition to optimize delivery channels. This is done by tapping and joining numerous data, including prior driving practice [20].

Automobile: Big Data has taken full command of the automobile business and is operating it evenly. Big Data drives the automobile business astonishingly and has provided never before results. Big Data has supported the automobile business to accomplish things beyond our intelligence [20]. From examining trends to interpreting stock chain administration, looking after our customers, and realizing our dreams of connected cars, Big Data is managing the automobile business well and truly insanely.

Telecom: Business is the heart of each digital innovation that is happening throughout the world. With the ever-gaining fame of smartphones, it has filled the telecom enterprise with huge volumes of data. And this data is like a gold mine, and telecom corporations require apprehension how to dig it. Through Big Data and Analytics, corporations can give clients seamless connectivity, thus eliminating all the network restrictions. Now with the guidance of Big Data and Analytics, companies can trace regions with low- and high-network traffic and, therefore, need to assure trouble-free network connectivity. Other enterprises, such as Big Data, have assisted the telecommunication industry to better comprehend its clients. The telecom industries are now offering customers as many customized offers as possible. Big Data is behind the data revolution we are currently experiencing [23, 24].

4 Conclusion and Future Scope

Big Data methods are invented to manage very huge and complicated datasets that cannot be processed by utilizing conventional systems. Originally, Big Data gained leverage to alter large volumes of automated data due to social media's bizarre extension. Since then, it has been utilized to process huge complicated datasets produced as a consequence of several experimental operations, construction methods, and network logs. In a very small period, Big Data has enrolled the technology platform and has established its presence in the field of technology and industry. Acknowledging the true potential of Big Data, companies are now passionately seeking to influence this technology to take advantage of the business and not face competitive risks in the long run. This chapter examines the notion of Big Data and the various technologies used to manage Big Data. For an industry that trades billions of dollars each year, Big Data is observed as a necessity rather than a luxury. It is no secret that Big Data is causing big changes in the business world. There are many advantages of Big Data, and it can be applied to regions that nobody thought of before.

Big Data is impacting the IT industry, just like some of the technologies that have been accomplished earlier. Extensive data produced from sensor-equipped devices, mobile phones, cloud computing, social media, and satellites can help various companies enhance their choice-making and drive their business to a different stage. Day after day data is produced so fast that traditional databases and other data storage systems gradually leave the storage, retrieve, and find relationships among data. The perplexity is that corporations require in-house skills and best methods. The downside of this is that Big Data has a service and advising boom.

The demand for solutions is so hot that all corporations are searching for a Big Data approach. Companies such as Google, Yahoo!, General Electric, Cornerstone, Microsoft, Kaggle, Facebook, Amazon are funding a lot in Big Data analysis and plans.

References

1. Gandomi, A., & Haider, M. (2015). Beyond the hype: Big Data concepts, methods, and analytics. *International Journal of Information Management, 35*, 137–144. https://doi.org/10.1016/j. ijinfomgt.2014.10.007
2. Liang Ting-Penga, & Liu Yu-Hsi. (2018). Research landscape of business intelligence and Big Data analytics: A bibliometrics study. *Expert Systems with Applications, 111*(30), 2–10.
3. Balakrishnan, S. (2019). An overview of agent based intelligent systems and its tools. *CSI Communications Magazine, 42*(10), 15–17.
4. Sun, Z., Sun, L., & Strang, K. (2016). Big Data analytics services for enhancing business intelligence. *The Journal of Computer Information Systems*, 1–8. https://doi.org/10.1080/0887441 7.2016.1220239
5. Normandeau, K. (2013). *Beyond volume. Variety and velocity is the issue of Big Data veracity.* Inside Big Data.
6. SAS Whitepaper. (2013). Big Data meets big data analytics. http://eric.univ-lyon2.fr/~ricco/ cours/slides/sources/big-data-meets-big-data-analytics-105777.pdf
7. Oussousa, A., Benjelloun, F.-Z., Lahcen, A., & Belfkih, S. (2017/2018, June/October). Big Data technologies: A survey. *Journal of King Saud University – Computer and Information Sciences, 30*(4), 431–448.
8. Watson, H. J. (2014). Tutorial: Big Data analytics: Concepts, technologies, and applications. *Communications of the Association for Information Systems, 34*, 1247–1268.
9. Storey, V. C., & Song, Y. (2017). Big Data technologies and management: What conceptual modeling can do. *Data & Knowledge Engineering, 108*, 50–67.
10. Ishwarappa, J. A. (2015). A brief introduction on Big Data 5Vs characteristics and hadoop technology. *Procedia Computer Science, 48*, 319–324. https://doi.org/10.1016/j.procs.2015.04.188
11. Yuri, D., Cees, D. L., & Peter, M. (2014). Defining architecture components of the Big Data ecosystem. In *Proceedings of 2014 international conference on collaboration technologies and systems (CTS)* (pp. 104–112).
12. Kumar, M., Punia, S., Thompson, S., Gopal, D., & Patan, R. (2020). Performance analysis of machine learning algorithms for Big Data classification. *International Journal of E-Health and Medical Communications (IJEHMC), 12*(4), 60–75.
13. Labrinidis, A., & Jagadish, H. V. (2012). Challenges and opportunities with Big Data. *Proceedings of the VLDB Endowment, 5*(12), 2032–2033.
14. Punia, S. K., Kumar, M., & Sharma, A. (2021). Intelligent data analysis with classical machine learning. In S. S. Dash, S. Das, & B. K. Panigrahi (Eds.), *Intelligent computing and applications* (Advances in intelligent systems and computing) (Vol. 1172). Springer. https://doi. org/10.1007/978-981-15-5566-4_71
15. A personal perspective on the origin(s) and development of "Big Data": The phenomenon, the term, and the discipline (Scholarly paper no. ID 2202843) Social Science Research Network (2012).
16. Yadav, S. P., Mahato, D. P., & Linh, N. T. D. (2020). *Distributed artificial intelligence: A modern approach* (1st ed.). CRC Press. https://doi.org/10.1201/9781003038467
17. Minelli, M., Chambers, M., & Dhiraj, A. (2013). *Big Data, big analytics: Emerging business intelligence and analytic trends for today's businesses.* (Chinese edition 2014). Elsevier.
18. Majumdar, J., Naraseeyappa, S., & Ankalaki, S. (2017). Analysis of agriculture data using data mining techniques: Application of Big Data. *Journal of Big Data, 20*, 1–15.
19. Yadav, S. P. (2020). Vision-based detection, tracking and classification of vehicles. *IEIE Transactions on Smart Processing and Computing, 9*(6), 427–434, SCOPUS, ISSN: 2287-5255. https://doi.org/10.5573/IEIESPC.2020.9.6.427
20. Habib ur Rehman, M., Ibrar, Y., Salah, K., Imran, M., & Jayaraman, P. P. (2019). The role of Big Data analytics in industrial internet of things. *Future Generation Computer Systems, 99*, 247–259.

21. Perera, C., Liu, C. H., Jayawardena, S., & Chen, M. (2014). A survey on internet of things from industrial market perspective. *IEEE Access, 2*, 1660–1679.
22. Yadav, S. P., Agrawal, K. K., Bhati, B. S., et al. (2020). Blockchain-based cryptocurrency regulation: An overview. *Computational Economics*. https://doi.org/10.1007/s10614-020-10050-0
23. Tabesh, P., Mousavidin, E., & Hasani, S. (2019). Implementing Big Data strategies: A managerial perspective. *Business Horizons, 62*(3), 347–358.
24. Xu, L. D., & Duan, L. (2019). Big Data for cyber physical systems in industry 4.0: a survey. In *Enterprise information systems, 2019 – Enterprise information systems*. Taylor & Francis.
25. Cirillo, D., & Valencia, A. (2019). Big Data analytics for personalized medicine. *Current Opinion in Biotechnology, 58*, 161–167.
26. Bhardwaj, A., Al-Turjman, F., Kumar, M., Stephan, T., & Mostarda, L. (2020). Capturing-the-invisible (CTI): Behavior-based attacks recognition in IoT-oriented industrial control systems. *IEEE Access, 8*, 104956–104966. https://doi.org/10.1109/ACCESS.2020.2998983
27. Shankar, A., Pandiaraja, P., Sumathi, K., Stephan, T., & Sharma, P. (2020). Privacy preserving E-voting cloud system based on ID based encryption. *Peer-to-Peer Networking and Applications*. https://doi.org/10.1007/s12083-020-00977-4
28. Kumar, S., et al. (2021). Forecasting major impacts of COVID-19 pandemic on country-driven sectors: Challenges, lessons, and future roadmap. *Personal and Ubiquitous Computing*. https://doi.org/10.1007/s00779-021-01530-7
29. Raheja, S., et al. (2021). Modeling and simulation of urban air quality with a 2-phase assessment technique. *Simulation Modelling Practice and Theory, 109*, 102281. https://doi.org/10.1016/j.simpat.2021.102281
30. Furht, B., & Villanustre, F. (2016). *Big Data technologies and applications*. Springer International Publishing. www.springer.com. https://www.springer.com/gp/book/9783319445489
31. Hung, P. C. K. (Ed.). (2016). *Big Data applications and use cases*. Springer International Publishing. www.springer.com. . https://www.springer.com/gp/book/9783319301440

Digital Marketing: Transforming the Management Practices

Priyanka Malik, Madhu Khurana, and Rohit Tanwar

1 Introduction

Digital Marketing is the marketing in which the products and the services are promoted and sold over online through the help of the Internet. The main promotion happens through an online medium, like mobile phones, digital devices and display promotion. Many companies are now operating through online communication where they promote their products on the online platforms, and this helps to gain more customer views, and also through the tools we can see the insights how many users or audience were covered in the online world. Digital marketing has the capability of engaging the customers [1]. Best-in-class digital marketers must learn how to apply artificial intelligence (AI) techniques to their digital marketing campaigns as AI has become more common in the digital marketing environment. In today's world, people opt for online healthcare systems [2], and at their comfortable times, they can get direct responses and connect with experts from their comfortable environment. By strengthening the exposure of the hospitality services in the online space [3], digital marketing has even infused new life into the healthcare industry. To attract the right audience, marketing companies use Big Data and the Internet of Things (IoT). Although IoT's [4] tireless expansion has been going on for decades,

P. Malik
Amity International Business School, Amity University, Noida, India
e-mail: pmalik2@amity.edu

M. Khurana
Gloucestershire College, Gloucester, UK
e-mail: madhu.khurana@gloscol.ac.uk

R. Tanwar (✉)
Department of Systemics, School of Computer Science, University of Petroleum and Energy Studies, Dehradun, India
e-mail: r.tanwar@ddn.upes.ac.in

© The Author(s), under exclusive license to Springer Nature Switzerland AG 2022
F. Al-Turjman et al. (eds.), *Transforming Management with AI, Big-Data, and IoT*, https://doi.org/10.1007/978-3-030-86749-2_6

its acceleration is staggering and much more visible in the digital marketing environment. Forecasting [5] is an essential aspect of the marketing planning process, both for annual expenditure and individual campaigns. Given the rise in digital marketing spending, it is now more important than ever to create more reliable estimates of digital marketing returns.

Digital marketing has some tools or methods such as Search Engine Optimization (SEO), content marketing, Search Engine Marketing (SEM), Artificial Intelligence, Programmatic Advertising, Chat Bots, Video Marketing, Social Messaging Apps [6], Visual Search and Social Media stories. Digital payment has also increased the demand of digital payment [7]. In addition, security issues [8] should be at the forefront of the minds of digital marketers who do not want their enterprise to reach an unfortunate endpoint.

Digital marketing is growing at a higher rate in this era as every company or business is promoting their services online only. Digital marketing is also known as Internet, Web or Online Marketing.

In today's era, digitalization has taken place where everything is digital. Many products can be promoted through digital marketing [9]. These tools have made life more easier for the small-medium enterprises where they can see the sales growth, the number of viewers, performance and budget. These tools help the company to make sure that it makes the profit and helps the customer in moderating the control of satisfaction. Traditional marketing is nowadays more backwards than modern marketing. In traditional marketing, only the basic methods are used to promote and sell the products and services.

Just like promotion of products on television, newspapers, radios, here only large audience is captured but the companies are not sure till where the audience would be captured.

In online marketing, things are different where we see that promotion of products and services are processed in the digital form; display is needed with Internet connectivity, hence, digitalization takes place. Online communication takes two side phases where social media and blogging comes in place. Social media marketing helps to promote the business by providing visual advertisement, banners and pop-ups.

Video marketing is also emerging in this sector where few seconds of video is made of the product where the ad is acted in such a way to attract the consumer. The visuals attract more attention. The more creative the company gets online, there are more chances to increase the sales and revenue for the company [10].

Medical device and diagnostic firms create some of the most technologically innovative devices on the market, but their marketing strategies often fall behind those of other high-tech industries. About the challenges of marketing in a heavily regulated world, medical device and diagnostic companies who offer automated diagnostic tools, as shown in [2], can differentiate themselves in their respective markets by implementing a strong digital marketing strategy.

1.1 Radial Benefit of Online Marketing

Public relations are increased due to the connectivity of the business online. This also creates the opportunity to work together in the corporate social responsibility and work for the environment.

- *Advertising*: It is defined as the medium where the companies can promote their business through online and offline methods. It is a paid source task in which visuals and information would be provided so that the customers can be attracted [11].
- *Connectivity*: While being online, using the Internet makes the business correlated to each other, and thus consumers and websites are connected to each other. The customer can login to the website any time and can gain information. It is the 24 × 7 available service. As shown by the authors [4], when smart cities become a reality, lightspeed connectivity and massive data provide digital marketers with numerous opportunities and a new set of challenges (Fig. 1).

- *Sales Promotion*: Sales promotion happens when there is a product launch and through visual marketing demand is created. After demand is created, the sales are promoted through digital marketing. It helps to generate more revenue and gets to know more insights in the dashboard of the company [12].

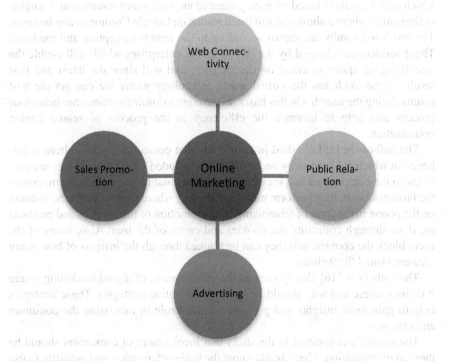

Fig. 1 The radial chart of digitization

1.2   Objectives

1. To analyse the effectiveness of online marketing as compared to traditional marketing tools
2. To analyse the role of digital marketing in improving the customer behaviour
3. To Identify the challenges faced by a marketer in this digital era
4. To understand the importance of online advertising in changing market scenarios

2   Literature Review

According to [13], the author explained that digital marketing helps to formulate the buying behaviour of the customer. This can track how many customers have visited the profile or website and where they have clicked more on the website. Clicking tool is available on the chrome extension, which when activated we can see the heat map on the screen. Heat map is the tool where three colours come: green, red, and yellow. The green colour shows the least clicked button on the website.The yellow colour shows mostly clicked buttons on the website.The red colour shows the highest number of clicks on the website.

P.K Kannan in [14] justified in his study that digital marketing has phases in which social media is based on user-generated engines which means search engine optimization where it shows the top listed results on Google Chrome or any browser. The top listed results are shown according to the search perception and methods. These services are charged by the digital online enterprises which will enable the search engine spider to crawl on the websites and will show the latest and best results at the top.It has the virtual world technology where we can get the best results during the search. On this basis, we also get to know the consumer behaviour process and help to increase the efficiency in the process of search engine optimization.

The authors in [15] clarified in their study that consumer digital culture is followed in which cyber security and e-office are included where the services are provided to the company so that they can protect their data and critical information of the business. Also, they perform mobile marketing where they promote the banners on the phone in the form of advertising. The promotion of the services and products are done through collecting the cookies and cache of the user. Also, many of the users block the cookies, still they can be tracked through the insights of how many viewers visited the website.

The authors in [16] clearly showed the effectiveness of digital marketing where it defines where and why should we apply these online strategies. These strategies help to gain more insights and perform a crucial role in increasing the consumer attractiveness.

The researchers justified in the study that involvement of consumers should be there in digitalization. They should know the basic information and usability to use the Internet and the websites. They should have the knowledge to operate the digital media. They should have the loyalty towards their preferred brand and make sure to respond well in the workshops held by the company. Still many of the customers believe in traditional or offline markets, but in this changing generation they should know that online sites are safe and protective in payment [17].

Anastasi Sotnikov in [18] said that it is important to analyse the tools and the methods of digital marketing that how much response it supports in the calculation of the impact of the sales. It also describes that at what time the marketing strategies should be used to enhance the business quality. The search engine optimization should be used at enough cost to gain the top search results for the company. Social media also comes in point where the company has to foresee the targets to complete and formulate the video blogging and banners on the website.

Charles Gibson enhanced the study that is important in this generation to build a website for a company so that the customers can know about the information of the products and services they provide. It must examine how many viewers visit their profile or website and where the heat map is included to know where the customer was engaged much in the website [19].

In [20], the author explained in the paper to maintain the brand reputation it is compulsory to build a strong social media platform where the content should be easy and understandable. There a lot of customers who get confused about what the company is actually selling. In the social media world, it all depends upon the video, content and visuals. Authors in [21] identified that it is necessary to analyse the use of digitalization in India. Many of the customers are unaware of the digital media. Due to the hype of digital India, customers only know that the generation has changed from offline to online but they do not have the access to the Internet or they still have the issue of connectivity.

3 Digital Marketing Generation

Digital marketing borders all the electronic devices or services into one platform. It puts all the marketing strategies into a digitalization. It is a platform where it concludes all the tools to enhance the business online. Digital marketing can also be called as online marketing communication [22].

Digital marketing consists of tools that help to promote the business online and gain more attractiveness and these are given below.

3.1 Categories/Tools of Digital Marketing

There are categories in digitalization that help to gain more followers and help to retain the customers in less time. These tools help the business in expanding over online and getting more insights towards the taste or preference of the customers. It also helps to give an impact towards the sales revenue which loses out in the traditional marketing. Automation is the biggest advantage to online marketing, where chatbots are used reducing the manpower and mistakes.

1. *Search Engine Optimization/ Search Engine Marketing (SEO/SEM)*: Search Engine Optimization and Search Engine Marketing are the tools of digital marketing where SEO makes sure that the number of results are more and makes sure while searching the keyword the brand results comes first in the search bar. For example, if we type only G as a keyword, it will autofill the link with Google. in; this is done by the online marketing and formulation of the paid search. The paid advertising is done with the search engine and shows the best result when paid higher (Fig. 2).

2. *Email Marketing*: Email marketing is used the most in this sector where a larger audience is captured. The invitation mails and proposals are sent to the higher management to opt for the digital services where these agencies can provide higher form of promotion in fewer budgets. Email marketing uses different kinds of tools, and they are as follows (Fig. 3):

- *HubSpot Sales*: This tool is also known as Sidekick which lets us open all the emails at the scheduled time and lets us know the contact history of the recipient. It also shows the social media handle of the user and all tweets it has done. It also shares the contact professional history and lets it make sure when to reply to all the emails' answers.

Fig. 2 Digital marketing tools

Fig. 3 Email marketing tools

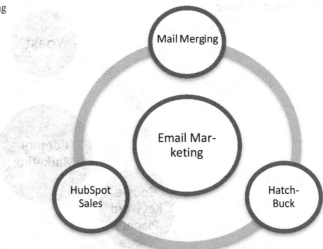

- *HatchBuck*: It helps to generate the leads, formulate new templates for the proposal and helps to improve client communication. It also tells the user activity and can judge when the user will be ready to buy the services.
- *Mail Merging*: Mail merging is the common tool where the recipients usernames or id's are saved at one place and when giving the proposals we just have to click the enlarge mailing preference contact which contains all the usernames, thus sending unlimited number of proposals to large audiences at one time.

3. *Content Marketing*: Content marketing is the form of promotion where the information is provided about the products and services online. It is the form where the customers can be attracted with the tag lines, one liners, quotes, blogs, hashtags, keywords etc. There are some categories under content marketing which helps to promote the business with more efficiency (Fig. 4).

These tools help to make sure that the content is on point and does not harm any kind of rules and regulations.

- *UberSuggest*: This tool helps to come up with new blog topics and with one keyword "suggest" and the related topic it will show an unlimited number of keywords which are offline and in current trend. We can also search in more depth using this tool under the chrome extension or we can google it from the search bar.
- *Google Keyword Planner*: This tool helps to find the potential keyword online and tells which are the most trending topics followed on the Internet. This will help to create the hype among the customers.
- *Yoast*: It is a plug-in used in the search engine software or Google chrome where it will give an extra head start for the user to put on the trending option, and thus the customers can read the blog and increase the higher ranking in the search engine.

Fig. 4 Content marketing

4. *Social Media Marketing*: Social media marketing is the biggest form of tool or category under digital marketing that helps to grow the business from low to high. It is a platform where almost every citizen of the country is available online. Through the help of social media applications, the company promotes its business in the form of banners, pop-ups, instant video ads, etc. Under SCM, there are some tools which help to promote the business with more efficiency and are as follows (Fig. 5):

- *Promo Republic*: It is a tool that runs the business or handles the company's social media on auto pilot. It includes scheduling, collaboration, monitoring, handling of posts, reporting to feedbacks. It handles all the facilities which manpower relies on. It saves up to 20 hours of working time.
- *Hoot Suite*: It is a kind of bot that schedules all the posts when to update and monitors the social media users account to stream the scheduled messages which should be replied to the customers.
- *MeetEdgar*: It is the best digital marketing tool that adds tweets, updates the post of LinkedIn and automatically uploads the post on Facebook into different categories according to the library list order. It also makes the calendar page to post the updates at a particular time.

5. *Marketing Automation*: Marketing automation has almost nailed every sector of business where it will automatically upload the blogs of the business and about the products and services. There are chat bots that will talk to the customer if it feels there are any problems regarding any situation or product.

Fig. 5 Social media
marketing

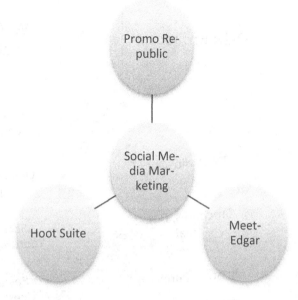

Optimize Press is also a tool under automation where it creates the pages on the blogs and will keep reminding to customize the template according to the taste of the company.

6. *Video Marketing*: Video marketing is a trend nowadays where the bloggers and even the business are uploading the videos of their products in an attractive way. The seconds videos make the video more pleasing, and visuals are more powerful than the content writing.

Buffer is an extension available to the chrome bar; when enabled it will automatically show the videos that are trending, and to make them trending search engine marketing takes place (Fig. 6 and Table 1).

Traditional marketing is less used in this era as the digital era has overtook the pace. After the new measures taken by the government, it has made everything online because now the company can actually analyse the data with the different tools available or made for the analysis.

Digital marketing is just the promotion of the products and services with more detailing, including digital media. For example, now we can measure the TRP of the media news through the measuring mergers and subscribers' limit.

As we can see, the balance of digital marketing is more and less of traditional marketing. It includes more of the advancement relating with the technology and has nearly killed the offline markets. According to DigitalSpace, before 2025, digital marketing will fully overtake the old broadcast marketing strategy.

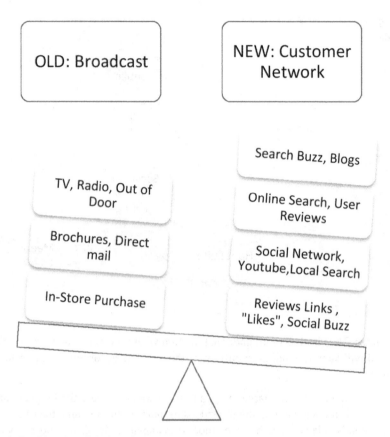

Fig. 6 Traditional vs digital marketing

Table 1 Traditional vs digital marketing

S. No.	Traditional marketing	Digital marketing
1.	It is more of limited space where only specific audience is captured.	It is a medium or platform where anyone can convey brand messages loud and clear.
2.	The audience needs to contact specially with the business by going to the customer care market.	The audience can directly interact with the business profiles online.
3.	Magazines, newspapers, broadcasts, postcards, telephone, sms etc. are used in traditional marketing.	Emails, blogs, automation chat bots, social media, tweets, LinkedIn updates, Facebook posts, YouTube etc. are used in online marketing.
4.	Low level of customer engagement is involved.	High level of engagement is involved.
5.	It is more expensive method to apply in the field.	It is less expensive method and enlarges the opportunities for the company to target.
6.	A large audience is captured, but it cannot be analysed or measured.	Same goes with the digital era but in this method we can measure the data with the tools.

4 Lead Generation

In the digital marketing world, lead generation should be enhanced so that the customers are engaged with the brand and follow the hype. It is the responsibility of the company to make sure that the customer should be retained and do not change the brands (Fig. 7).

- Firstly, the company is totally a stranger to the audience or the customer. Through blogs, social media and keyword pages, the audience would be attracted.
- Calls to action, landing pages, forms, contracts are the tools from where the customers would be visiting the websites and would be educated and engaged.
- The leads mean forming a relationship that will work in the forms of emails, workflows of CRM generation.
- To create an advocacy, social media, smart call-to-action and email workflows will help to form more customers and will help to promote the business.

There is a total transparency in the lead generation so that there is retention of the customers in the best way of formulation. The company formulates the web search and content marketing and interacts with the customers in the best possible way.

Lead generation is formed by various sources or activities that are conducted by the samples or designs made by the company via the Internet. The lead is generated

Fig. 7 Lead generation

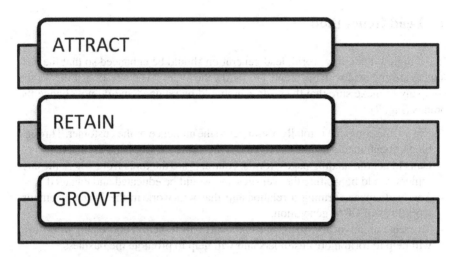

Fig. 8 The marketing concept

by the salesperson of the company and makes the generated value by presentations and showing the strengths of the company.

The lead generation is conducted in the search engine optimization where the company can enhance the web results and increase more traffic on the website. The website helps to generate more profit, and through the content and heat map it helps to increase the viewer's response and conduct a sales report (Fig. 8).

- *Attract*: In the first stage, it is defined that the customers should be more attracted towards the product and service through mainly e-commerce portal or websites.
- *Retain*: In the second stage, it is said that retaining the customers is more important and less expensive than making new customers. To retain the customers, more offers and discounts should be provided on the new technology products or services so that the loyalty is maintained for the customers.
- *Growth*: In the third and final stage, growth is all it matters where bringing out new technology or ideas so that there is development and growth and can compete with other company competition.

5 Challenges in Digital Marketing

- *Segmenting the Audience*: Demographics, lifestyle and behaviour are the factors that are difficult to analyse in the digital marketing world. These factors makes more complicated in finding out exact data. The data is mostly taken in tentative way where formulating the actual size or sample is not desirable. There are tools to make the data right but that needs more budget. For the small-scale companies, it is difficult to calculate the actual size data and take out the segment audience where to target.

- *Social Specific*: In the digital marketing world, it is all dependent on social media. Everything is based on social and digital media, and it costs a lot of money. Budget is a big concern and a big challenge for the marketing agency. The agencies need a lot of funds to promote the products and services online.
- *Invest in Locals*: Investing in locals means not always investing in Google business but also the chat bots' questions and answers lead to some good results. But Google is the key to enhance all the possible outcomes to overcome the marketing promotions and web search. The search engine optimization works upon the Google results and that costs a huge amount of money.
- *Paid Advertisements*: Budget is the biggest issue in digital marketing; these tools cost huge amount of funds to promote the products and services online. The promotions are done so that the customers can be retained and attracted towards the products. But there is a flaw that budget needs to be big which many companies fail to figure it out.
- *Design Developers*: To build a website, a developer is needed so that the best website can be developed and more followers can be gained. But the developer charges a huge amount of money to handle the server and website so that there is no lag for the consumer when they use the website. The developer should have the best knowledge so that there is efficiency on the website. UFX and VFX tools are the most expensive software that are used by the developers.
- *Challenge in Creating Content*: One of the biggest challenges is creating the best content in the digital marketing. The customers want the short and best banners and should be catchy at first sight. To create this, intelligence and skills are needed.
- *Challenge in Promoting Content*
  - *Social Media*: In promoting the business in social media, it costs more money and promotion time. The promotional activity consists of a huge amount of time and takes a lot of content. The Facebook ads and Twitter tweets are difficult to manage in this field. The promotional activity on the Google also takes budget. The keywords are bought by Google so that the search engine optimization and search engine marketing hit the customer in rightful way.
  - *Email*: Email marketing consists of lot challenges where the response is not tracked by the company. The emails are sent in bulk and are not tracked by any of the tools. The emails can be undo while sending to the recipients, but it cannot be sure that the replies would be back. The proper proposals are sent to the customers, but the forms are not always accepted.
  - *Generating New Leads*: Without the contacts, it is almost impossible to market the products and find the clients. So approaching new leads is very difficult, and without that, the business would be difficult to run, that is, the biggest challenge is generating the leads. The sales person tries to find out the potential in the market, which is the challenge in digital marketing.

6 Research Methodology

In this chapter, according to the objectives, the following particulars were performed and finalized by the review of literature, by experience and practical-based conclusions. The research is being performed on the experience-based study. It is completed on the basis of all the experiences gathered through practical and theoretical knowledge gained during the internship in ADG Online Pvt Ltd, a digital marketing firm.

Secondary data collection: All the theoretical knowledge was gained during the course of my internship, for example, using different tools and formulating in lead generation and also formulating the new content for the products and services and promoting in social media like LinkedIn, Facebook, etc.

According to the first objective, we have analysed through charts and personal experience that nowadays digital media has taken over the traditional method and has increased the pace for the customers to interact with the company giving the 24 × 7 service through digital media.

According to the second objective, we have analysed that being online for a company, it is easy for the customer to interact with them and can access the website and other online services anytime. This has made them more reliable and efficient in all the ways to conclude the services for them at any time.

According to the third objective, we have analysed that being a digital marketer there are a lot of challenges to be faced. The online world seems to be easy, but working with it is more difficult. The digital marketing takes a lot of skills and intelligence. The marketer should be more creative and confident over its strategies to use in the digital world.

According to the fourth objective, we have analysed the importance of online marketing on the above comparison between traditional and digital marketing. It captures more of audience and that can be calculated with the tools. Online advertisements make the customer more aware at a fast pace because in the current generation all of the human beings are connected to Internet so that it makes easy to target the potential customers.

The survey questionnaire was refined, and a total of 51 respondents filled it. The survey took place with the help of online forms and was filled by the audience online. The paper consists of independent variables such as tools of digital marketing and web learning and challenges of it, and there were dependent variables such as age, gender, income and educational qualifications.

7 Recommendations

Digital marketing is evolving everyday in this generation. With the challenges, there are some recommendations that will help the audience to know where they can prevent the cause and do not face any difficulty in providing the services to the company.

- There should be a proper knowledge of the tools that are used in digital marketing. It is not only used in this field but it includes different aspects which needs specialization on using those tools.
- Focus on only one tool for the promotional marketing, as getting or working under various different marketing will lead to more chaos and will lose more clients at one time.
- The blogging and calendar pages are different; mostly many times they look similar but they are different. The blogging contains the content of the products and services, but the calendar page tells about the company day-to-day activity what they have achieved or what their interests are.
- Websites include bad links; learning web programming will give a proper exposure where you will automatically find out the cons in the website.
- If working in e-commerce portal, remember to use the heat map where you can analyse where your customer scrolls on the website.

8 Conclusion

Digital marketing makes sure that all the products and services are promoted on the digital media with no interruptions, and can capture a huge number of audiences at once. Not only capturing, it also helps to provide the services any time of the day.

Modern technology has changed a lot. Besides the growth of modern technology, the technique of businesses has also changed. In this regard, digital marketing takes the leading position. The present modern generation does not want to go to any shop or shopping mall to buy anything. They want to buy everything online. So, the various companies always try to stay connected with their consumers or customers via the Internet.

Direct advertising makes it easier for customers to do business. Through the websites and social media, easy brand promotion makes the companies more profitable. Our study shows the challenges of digital marketing and how does it impact consumer behaviour. The digital marketing gives the customers' feedback option so that the company can change the requirements according to their needs and uses, that is why digital marketing, despite the challenges, is famous at present.

Since there are many opportunities in this sector, there is a large scope for the youth where they can learn more about promotion and increase their skills in the jobs. Many companies are hiring candidates who are specialized in this sector.

Annexure 1: Questionnaire and Responses

Gender:
- Male:

- Female:

Gender
51 responses

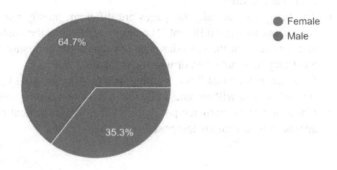

Age:
- 21–30
- 31–40
- 41–50
- 51–60

Age
51 responses

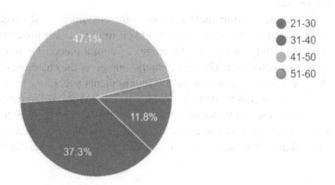

1. Marital status:

   - Single
   - Married

Marital Status

51 responses

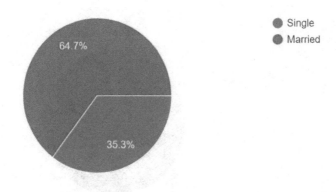

2. Life stage:

- Young and dependent
- Young and independent
- Less young and independent
- Family with children at home
- Empty nesters/retired

Life Stage

51 responses

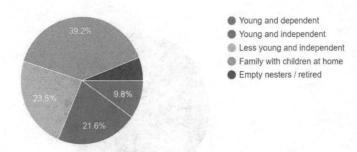

3. Type of customers:

- Student
- Professional
- Salaried
- Business
- Housewife

- Retired

Occupation

51 responses

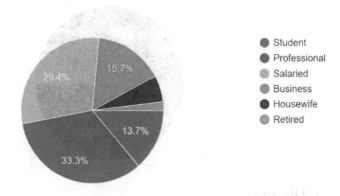

Student
Professional
Salaried
Business
Housewife
Retired

15.7%
29.4%
13.7%
33.3%

4. Qualification:

- U.G
- P.G
- Professional
- Others

Qualifications

50 responses

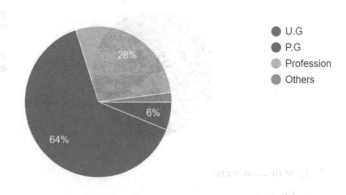

U.G
P.G
Profession
Others

28%
6%
64%

5. Gross annual income in INR:

- Less than 4 lacs
- 4–5 lacs
- 5–7 lacs

- Above 7 lacs

Gross Annual Income

51 responses

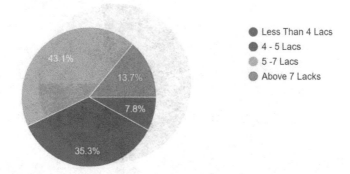

- Less Than 4 Lacs
- 4 - 5 Lacs
- 5 -7 Lacs
- Above 7 Lacks

6. Tax payer:
   - Yes
   - No

Tax Payers

51 responses

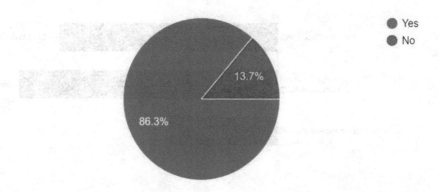

- Yes
- No

7. Which form of marketing is best?
   - PPC
   - SEO
   - SMM

Which Form of Marketing is Best?

50 responses

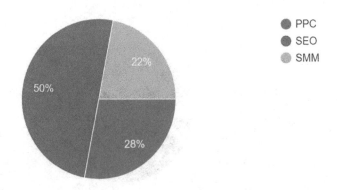

8. Which bidding options you use for PPC?

- Cost Per Click (CPC)
- Cost Per Thousand Impressions (CPI)
- Cost Per Action (CPA)

Which Bidding Options You use for PPC?

51 responses

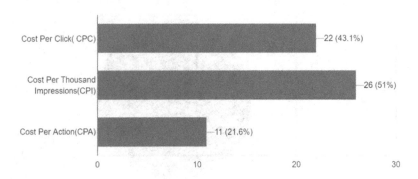

9. What are the traits of bad links?

(Any backlink that disturbs the guidelines of Google is called as bad link)

- Links from those websites that are not in Google Index
- Spam links from articles or blogs
- Poor authority website
- Penalized website

- Links from low traffic

What are the traits of bad links?

51 responses

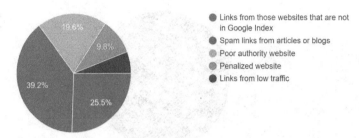

Legend:
- Links from those websites that are not in Google Index
- Spam links from articles or blogs
- Poor authority website
- Penalized website
- Links from low traffic

10. Does lead generation make your business easy?

- Yes
- No

Does lead generation makes your business Easy?

51 responses

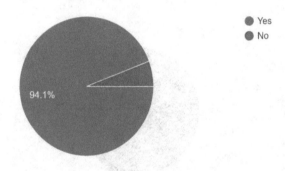

- Yes
- No

11. Do SEO/ SEM tools make your business successful?

- Yes
- No

Do SEO/ SEM tools makes your business successful?

51 responses

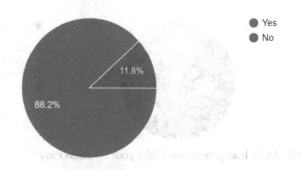

12. Using social media tools let your business promote your products?

- Yes
- No

Using Social Media Tools let your Business promote your products??

51 responses

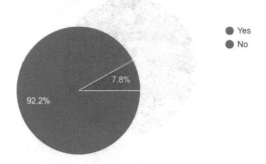

13. Which of the following tools you use for social media marketing?

- Tail Wind
- Sprout Social
- MeetEdgar
- Others

Which of the following tools you use for Social Media Marekting?

49 responses

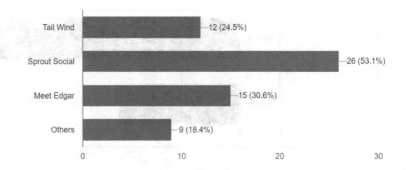

14. Is content writing important in digital marketing?

- Yes
- No

Is content Writing Important in Digital Marketing?

51 responses

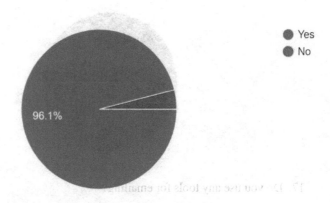

15. If Yes? Which tools you use for managing the content for products?

- Canva
- Animoto

If Yes? Which tools you use for managing the Content for Products?

51 responses

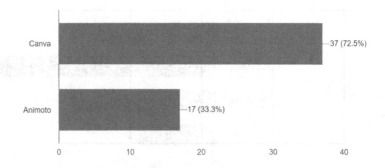

16. Is email marketing beneficial for your business?

    • Yes
    • No

Is Email Marketing Beneficial for your Business?

51 responses

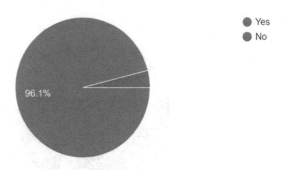

17. Do you use any tools for emailing?

    • Yes
    • No

Do You use any Tools for Emailling?

51 responses

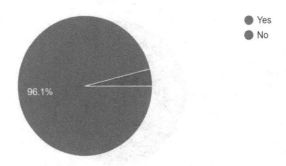

18. If Yes? Which ones?

- Mail Merging
- Hatchbuck
- Active Campaign

If Yes? Which One's?

49 responses

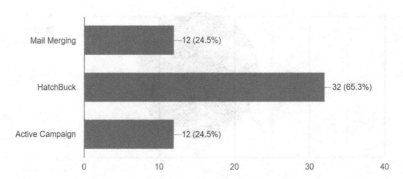

19. Being digital marketer, promoting products online is easy?

- Yes
- No

Being a Digital Marketer, Promoting Products online is Easy?

51 responses

20. Is it important to learn web programming in digital marketing world?

   - Yes
   - No

Is it Important to Learn Web Programming in Digital Marketing World?

51 responses

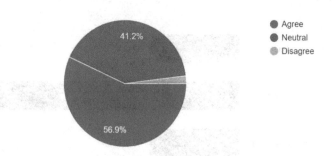

References

1. Smith, T. M. (2016). *Consumer perceptions of a brand's social media Marketing.* The University of Tennessee, Knoxville. Retrieved from http://trace.tennessee.edu/cgi/viewcontent.cgi?article=4332&context=utk_gradthes on 27 Mar 2016.
2. Punitha, S., Al-Turjman, F., & Stephan, T. (2021). An automated breast cancer diagnosis using feature selection and parameter optimization in ANN. *Computers & Electrical Engineering, 90*, 106958. https://doi.org/10.1016/j.compeleceng.2020.106958

3. Kumar, M., Alshehri, M., AlGhamdi, R., Sharma, P., & Deep, V. (2020). A DE-ANN inspired skin cancer detection approach using fuzzy C-means clustering. *Mobile Networks and Applications, 25*(4), 1319–1329. https://doi.org/10.1007/s11036-020-01550-2
4. Chithaluru, P., Al-Turjman, F., Kumar, M., & Stephan, T. (2020). I-AREOR: An energy-balanced clustering protocol for implementing green IoT in smart cities. *Sustainable Cities and Society*, 102254. https://doi.org/10.1016/j.scs.2020.102254
5. Kumar, S., Viral, R., Deep, V., Sharma, P., Kumar, M., Mahmud, M., & Stephan, T. (2021). Forecasting major impacts of COVID-19 pandemic on country-driven sectors: Challenges, lessons, and future roadmap. *Personal and Ubiquitous Computing.*
6. Kaiser, M. S., Mahmud, M., Noor, M. B. T., Zenia, N. Z., Mamun, S. A., Mahmud, K. M. A., Azad, S., Aradhya, V. N. M., Punitha, S., Stephan, T., Kannan, R., Hanif, M., Sharmeen, T., Chen, T., & Hussain, A. (2021). iWorkSafe: Towards healthy workplaces during COVID-19 with an intelligent pHealth app for industrial settings. *IEEE Access*, 1–1. https://doi.org/10.1109/ACCESS.2021.3050193.
7. Malik, P., et al. (2017). Consumer awareness of digital payment with special reference to the village area. *Pertanika Journal of Social Sciences and Humanities, 25*(4), 1585–1600.
8. Bhardwaj, A., Al-Turjman, F., Kumar, M., Stephan, T., & Mostarda, L. (2020). Capturing-the-invisible (CTI): Behavior-based attacks recognition in IoT-oriented industrial control Systems. *IEEE Access*, 1–1. https://doi.org/10.1109/ACCESS.2020.2998983
9. Sahai, S., et al. (2018). Role of social media optimization in digital marketing with special reference to Trupay. *International Journal of Engineering and Technology (UAE), 7*(2), 52–57.
10. Pawar, A. V. (2014). *Study of the effectiveness of online marketing on integrated marketing communication.* DY Patil University.
11. Ahmed, I. (2016). The effects of digital marketing on customer relationships in Bangladesh. Internship report, *BRAC Business School* (pp. 1–65).
12. Hajli, M. N. (2014). A study of the impact of social media on consumers. *International Journal of Market Research, 56*(3), 387–404.
13. Srivasankan, S. (2017). Digital marketing and its impact on buying behavior of youth. *International Journal of Research in Management & Business Studies, 4*(3) (SPL 1), 35–39.
14. Kannan, P. K., & Li, A. (2017). Digital marketing: A framework, review and research agenda. *International Journal of Research in Marketing, 34*(1), 22–45.
15. Stephen, A. T. (2016). The role of digital and social media marketing in consumer behaviour. *Current Opinion in Psychology, 10*, 17–21.
16. Yasmin, A., et al. (2015). Effectiveness of digital Marketing in the challenging age: An empirical study. *International Journal of Management Science and Business Administration, 1*(5), 69–80.
17. Marisavo, M., et al. (2007). An empirical study of the drivers of consumer acceptance of mobile advertising. *Journal of Interactive Advertising, 7*(2), 41–50.
18. Sotnikova, A. (2016). The application of digital marketing strategies to increase profits of the organization. Bachelor's thesis; Hame University of Applied Sciences.
19. Gibson, C. (2018). The Most effective digital marketing strategies & approaches: A review of literature. *International Journal of Scientific and Research Publications, 8*(2), 12–16.
20. Leeflang, P. S. H., et al. (2014). Challenges and solutions for marketing in a digital era. *European Management Journal, 32*(1), 1–12.
21. Singh, S. N., et al. (2016). Digital marketing: Necessity & key strategies to succeed in current era. Bachelor's thesis; Hame University of Applied Sciences.
22. Daiana, M. T., et al. (2015). A conceptual framework of the impact of social media marketing on consumer's relationship. *Recent advances on business, economics and development.* http://www.wseas.us/e-library/conferences/2015/Budapest/AEBD/AEBD-00.pdf

Real-Time Parking Space Detection and Management with Artificial Intelligence and Deep Learning System

Shweta Shukla, Rishabh Gupta, Sarthik Garg, Samarpan Harit, and Rijwan Khan

1 Introduction

Parking space management is one of the biggest problems faced in urban areas in both developed as well as developing countries. The number of cars running on the road is increasing with the car demand day by day, but the availability of parking areas does not meet the requirements. Along with this, the improper management of available parking spaces makes this problem worse.

There are many existing solutions to deal with this problem, such as barriers and ground sensors placed on every parking space, but they are expensive to use as they include installation cost, maintenance cost, and hardware cost and mostly not feasible in every scenario. Our approach is advantageous over the existing solution, which can monitor several parking spaces simultaneously reducing the cost [1].

The video to monitor parking areas is not new; see for the instance [2]. Many techniques proposed are not generalized for multiple types of parking lots. However, authors are proposing solutions that can be used in multiple scenarios like open parking space, off-street parking spaces, etc. Our solution not only converses the problem of parking but aims to solve almost every problem likely to occur in parking areas.

Our solution has three major parts in it, which builds up a complete system used in real time in parking spaces. Since authors are using live video streams, the capture can not only be used for parking space management but also the anomaly activities (like accidents, fire, burglary, etc.) which are captured in the camera. It will make the system splattered for parking occupancy detection, Parking space

S. Shukla · R. Gupta · S. Garg · S. Harit · R. Khan (✉)
Department of Computer Science and Engineering, ABES Institute of Technology, AKTU, Ghaziabad, UP, India

F. Al-Turjman et al. (eds.), *Transforming Management with AI, Big-Data, and IoT*, https://doi.org/10.1007/978-3-030-86749-2_7

127

prediction, and anomaly action detection creating a fully fledged application that can be used in real-life scenarios.

1.1 Parking Occupancy Detection

In parking space detection, the use of CNN and LSTM models is producing quite accurate results even with the presence of disturbances created by the partial occlusions, obstacles, shadows, different light conditions, etc. trained and tested upon the CNRPark dataset [1, 2]. The CNN model will produce output free or occupied based on a trained dataset. The one with no car in the picture will be labelled as 'free' and the one with a car in a slot will be labelled as 'occupied'. In the parking lot, each camera is assumed to be still in its position and hence the slots in the frames that are extracted will be fed to the model and the expected result will be attained.

1.2 Car Parking Availability Prediction Service Using LSTM

There are times when one can need information about the available parking space in a particular time period. It can be quite challenging to predict available parking space at a particular time before the head because parking space depends on many factors like place, time, weather, day, occasions, holidays, etc.

Since it is a time series-based problem, this can be solved using the Recurrent Neural Network (RNN) and Long Short-Term Memory (LSTM) model. LSTM is a type of RNN which is capable of learning long-time-based data [3, 4].

Since the factors under consideration will be different for each parking area, there will be a need for a unique LSTM model to be trained for each parking space. The dataset to be used to train these models will be collected over a period of time for each parking space using our real-time parking space detection system. Once good time-series data is observed, it will be ready to be trained and used [15].

1.3 Additional Anomaly Detection Using Deep Learning

Surveillance cameras are available at almost all public places, for example, shopping complex parking, banks, streets, which can be used in both parking space management and public safety as well. Authors can use computer vision capabilities to take small steps towards managing all these anomalous activities which may occur in parking spaces. Anomalous events, for example, violence or accident or any theft in a parking area can be detected by developing algorithms using deep learning.

To tackle anomalous events like accidents and violence in parking areas, authors have developed a Multiple Instance Learning (MIL) solution by holding only weakly labelled training videos. MIL is a type of supervised learning where instead of using every labelled instance, the author feeds a set of labelled instances called bags to a learner. This will help a deep learning model to learn anomaly scores for video segments [4, 5].

2 Related Work

The existing work on smart parking systems consists of combinations of hardware and software collaborations. The solutions that have been proposed till now to be addressed as smart parking systems are mainly focused on three types: Internet of Things (IoT)-based, networking protocol-based and software-based. The IoT provides a wide range of solutions to parking-related problems [16]. It involves usage of different types of sensors as they are the most important component of IoT-based smart parking systems as they play a significant role in collecting data. Types of sensors used in these systems are mainly Camera, Ultrasonic, Infrared, Cellular sensors, Radar and many others [17]. The usage of sensors and other IoT technology is based on factors such as ease of installation, usage of sensors per slot, detection autonomy, etc. These factors play a role in sensor selection strategy. Sensors used have the ability to detect the occupancy status of the slot in which they are installed. Sometimes, usage of sensors results in a complex infrastructure that involves structured cabling, electrical connections, aerial hindrances, etc. This makes the system 'not easy to maintain and increases maintenance cost'. Apart from above-listed sensors, there is RFID; it is an invasive sensor, as a card which is given to a user or installed on a vehicle, so whenever the user parks the car in a slot, it is determined if it is free or not. Use of radar sensors is another way to deal with the problem; these sensors use an algorithm to calculate the presence of any object in parking space and analyse how big it is and how it is positioned. This is then further analysed using software combinations.

Networking protocols are used along with sensors to pass information from sensors to smart parking systems so as to prevent the overhead of retrieving data from each sensor installed. Networking in smart parking has given better results, reliable throughput, less energy consumption and minimum delay. But still as the whole system relies on sensors and any fault in these will create a hindrance in the entire system. As per selected research papers, 40% of sensor-based solution did not specify usage of sensor networking in proposed solutions; among the rest, 85% have proposed usage of wireless implementation using Wi-Fi, Bluetooth and other wireless IoT protocols for communication, and remaining are using wired technologies such as Ethernet, USB, serial communications in user networks. Most of the solutions preferred wireless solutions because of reduced cost of deploying architecture [28, 29]. Improvements in networking protocols are also proposed for better performance [27].

Software solutions are information management systems mainly used for processing the accumulated data from sensors. To achieve the objective of smart parking and management systems, solutions are integrated with software technologies in order to provide intuitive tools which are easy to use and affordable for users [20]. Software solutions implemented provide features such as information management, analysis and prediction, and e-parking services. Some authors proposed a processed system in which driver requests and real-time information of the driver is collected, and slots are allocated through Variable Message Signs (VMS) or Internet [30]. A similar interface is provided to users, which allows to book parking space in some systems [31]. In both, the system directly communicates with sensors and processes data in real time. Many systems are handling states of parking slots: free or occupied. Some monitor and handle scenarios like illegal parking manners, unpaid parking and also monitor parking limits are exceeded or not [32].

Some solutions shows real-time parking information to the users and handle queries; payment information is also saved [33, 34]. Schema is also being used to store user location, payment details as shown in [35]. Many emphasise on cloud storage of data collected [23, 25].

Prediction using the collected information through sensors is another approach. This allows pre-booking of slots in nearby locations by taking other factors such as free space, distance, traffic flow [30, 36].

E-parking allows users to decide manual or automatic allocation of parking slots [37]. Once information is collected, it is being presented to the user through user side applications such as mobile or web interfaces.

3   Proposed Method

In this project, authors have proposed a way to reduce the congestion caused by the improper use of parking spaces. A real-time indication of occupancy of parking spaces by building a deep learning-based framework will be a robust, inexpensive solution to the problem. Here, authors are assuming parking spaces are preinstalled with surveillance cameras, from which the author will be taking live footage and feeding them to the frame extraction system followed by the slot extraction system. The extracted slots will be classified as free or occupied spaces, and this result will be live updated on the hosted server. The server is connected to mobile and web applications. The results of the CNN model will be collected over a long period and fed to the LSTM model which will give predictions for space availability in the parking area, considering the factors of day, time, weather conditions.

Initially, the model may or may not provide very accurate results, but when a large dataset is collected over a long period with large variations in considered factors, surely there will be an output leading to good results and dependency.

Along with the CNN model, the live stream frames are also fed into our anomaly detection model, to detect any anomaly that has occurred in the parking space. Some common anomalies that occur in parking space are Theft, Accident, Explosion, and Fights. Our model will be able to detect all these anomalies so that the admin can be notified about these activities in real time and he can take serious action accordingly.

3.1 Parking Occupancy Detection

Dataset used for training and validating the data:

The authors have used the CNRPark datasets for training and validating the Convolutional Neural Network (CNN) model. CNRPark+EXT is a dataset used in our smart parking system with roughly 150,000 images labelled; CNR-EXT, which is composed of images collected over the period of November 2015 to February 2016 where multiple weather conditions are covered by nine different cameras with different perspectives. This subset captures different situations based on light conditions and also includes 'partial occlusion patterns' due to obstacles such as trees, lampposts, etc. [8]. From this huge dataset, the authors used 70% of data in training and 30% for validating our results (Fig. 1).

Assessment of the Proposed CNNs

This chapter is targeted to find the occupancy of parking spaces using the frames obtained from the surveillance camera feed. The camera-based system is considered a more cost-effective approach as compared to other solutions implemented. The proposed Sequential Convolutional Neural Network Architecture consists of three convolutional layers. All the patches are shuffled and resized to a fixed size of 224×224 pixels. Hence, each image is classified independently from each other. In order to limit the problem of over fitting, authors split the dataset into 80% and 20% for training the model to classify whether the slot is empty or occupied and for testing, respectively. From the training dataset, 15% is used for validation. This allowed

Fig. 1 Different scenarios for space detection

us to make the model robust for the proposed solution to possible changes that may occur in the real-time situation [9, 10].

The authors trained the model with randomly cropped images to 224 × 224 × 3, while the testing is done with resizing them to 224 × 224 × 3 centrally. The hyper parameter 'relu' (rectified linear unit) is used as activation for all the layers and 'softmax' is used as activation in the output layer.

The first convolutional layer consists of 1,50,528 neurons (i.e. input shape of 224 × 224 × 3) with strides of 4 × 4 and kernel size 11 × 11 followed by the Max pooling and strides of size 2 × 2. The second convolutional layer consists of strides of size 1 × 1 with 20 filters and kernel size 11 × 11 followed by Max pooling and strides of 2 × 2.

The third convolutional layer consists of 30 filters and strides 1 × 1 and kernel size 3 × 3 followed by Max pooling and strides of 2 × 2. Hence, giving a total 46,951 of total trainable parameters (Table 1).

This model reaches an accuracy of 97.89. The accuracy may get affected by the change in factors like the height of the camera, clarity of images, angle of view, distance from parking space, etc.

Table 1 Assessment of the proposed CNNs

Layer (type)	Output shape	Parm #
Conv2d_1 (Conv2D)	(None, 38, 38, 16)	5824
Activation_1 (Activation)	(None, 38, 38, 16)	0
Max_pooling2d_1 (Maxpooling2)	(None, 19, 19, 16)	0
Conv2d_2 (Conv2D)	(None, 9, 9, 20)	38,740
Activation_2 (Activation)	(None, 9, 9, 20)	0
Max_pooling2d_2 (Maxpooling2)	(None, 4, 4, 20)	0
Conv2d_3 (Conv2D)	(None, 2,2, 10)	1810
Activation_3 (Activation)	(None, 2, 2, 10)	0
Max_pooling2d_3 (Maxpooling2)	(None, 1, 1, 10)	0
Flatten_1 (Flatten)	(None, 10)	0
Dense_1 (Dense)	(None, 48)	528
Activation_4 (Activation)	(None, 48)	0
Dropout_1 (Dropout)	(None, 48)	0
Dense_2 (Dense)	(None, 1)	49
Activation_5 (Activation)	(None, 1)	0

Total params: 46,951
Trainable params: 46,951
Non-trainable params: 0

3.2 LSTM Model for Parking Space Prediction

The proposed solution involves the use of Long Short Term Memory (LSTM), which is a specific type of Recurrent Neural Network (RNN) architecture. LSTM is able to learn long time series-based data more accurately by understanding the factors like time and date of visitors in a particular parking area. Other factors can be taken into consideration such as weather, occasion, day, holidays, etc. for better results [7].

3.2.1 Dataset

The data for this model will be generated over time as our system will be used. The collected data will be parking lot occupancy on a particular day which can be further characterized based on date, month, week, occasion, and weather [6].

3.2.2 Model Structure

A typical LSTM network consists of different memory blocks called cells (the rectangles in Fig. 2). Hidden states and cell states are the two states that are being transferred to the next cells.

Three types of gates are used in memory blocks:

(1) Forget Gate, (2) Input Gate, (3) Output gate [8]: These gates help in regulations and changes in the memory blocks responsible for remembering.

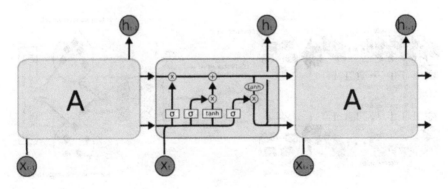

Fig. 2 Model structure

3.3 Deep Learning Model for Anomaly Detection

As authors are making a manless parking management system, authors need to improve the security also. For improving security, authors are implementing an anomaly detection system that will create an alert if anomalies like violence, theft, accident, etc. occur in the area covered by surveillance cameras. The frames from the video will be segmented and classified into two groups positive (for anomalous) and negative (for normal). Then, will make two instance bags, using this model will be trained and used.

3.3.1 Dataset

Authors briefly review the anomaly detection datasets which include long untrimmed surveillance videos that cover 13 real-world anomalies such as Accident, Stealing, Fighting, Explosion, and many more.

3.3.2 Model

The above dataset is used to identify the abnormal activities in the parking areas. To recognize the activities, it identifies the task in two outputs based on fourfold cross-validation (Fig. 3).

It constructs a 4096-D feature vector by averaging C3D [21] features from each 16-frames clip followed by an L2-normalization. The feature vector is used as an input for the nearest neighbour classifier. The Tube Convolutional Neural Network (TCNN) [11, 12] introduces the tube of interest (TOL) pooling layer to replace the fifth 3d-max-pooling layer in the C3D pipeline. The TOL pooling layer aggregates features from all the clips and outputs one feature vector for an entire video [13].

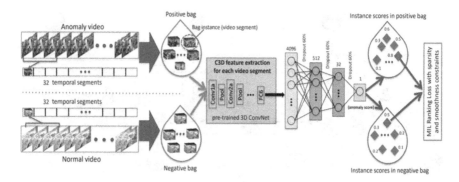

Fig. 3 Deep learning model for anomaly detection

3.4   User End Application

The final objective of our smart parking system is to provide a convenient and intuitive system for our users so to interact with our system effortlessly with the technology affordable to them for use.

Here, the author wants to provide a tool, as easy to operate as it could be and provide means to find the most convenient parking slot at the given time and location in the least time possible. The mobile technologies are growing day by day, reaching to every person, providing them an access to the new world of possibilities. Given the rise of this technology, the best solution was to build an Android mobile application accompanied with a web application which can provide our customer a platform through which they can easily either check or book the available slots for the location they are interested in. On top of the native Android language, that is, Java, authors have used Firebase to accompany us on the journey to be fast and easily accessible. Firebase helps to notify the updates in real time to the user. Authors have also integrated Google Maps in our application to allow users to search the location of the parking space directly from the Maps UI [14].

Our application starts with the Login screen where the user is asked to Login to our platform or continue as the guest user. Then the user is greeted with the welcome screen where he/she can search a location either by typing it in the search box or adding a marker on the Map which can be opened from the Map Icon besides the search button. Given the appropriate input from the user, authors display the details of the location, that is, Open Time, Crowd Prediction for that particular day and number of slots available in that parking space. Then the user has an option to book one or multiple slots for a specific time, also chosen by the user. As the user progresses, they will be sent to the payment gateway as they proceed to pay in the advance. After the payment is successful, they will be sent a confirmation through SMS and email and also can view their booking status through the Booking History menu. In the Booking History menu, the user can view their booking status, details and can edit the time or date of their booking. They will also be able cancel the request for a refund.

As the user interacts with the Android mobile application, the information about the user's booking history will get synchronised with the Firebase through Firebase API available for Android mobile application. As the user information gets synchronized with the Firebase DB, he/she can view their information about the booking status and history as a web application also (Fig. 4).

4   Results and Conclusion

A centralized and efficient image-based system is designed and developed for the management of parking spaces using deep learning models. The proposed smart system is equipped with cameras that are usually installed in the parking areas

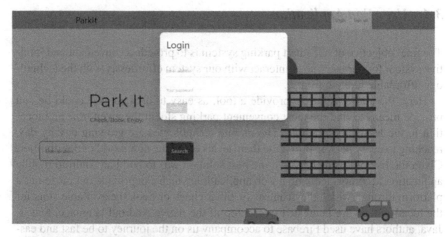

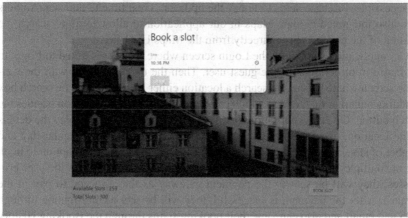

Fig. 4 Demonstration of user interface

which exempts the additional cost of installation and makes the system dynamic and easy to install. The overall performance of the CNN model used for occupancy of parking space in different weather conditions (sunny, cloudy, rainy) is reported as approximately 97.5 ± 0.25. The three performance measures: Accuracy, Recall and Precision are used for the classifiers for the various weather conditions, and they lie very similar in the range of 97.34–97.75%, 97.53–97.71%, and 97.27–97.67%, respectively (Fig. 5).

The proposed image-based system is suitable for real-time applications as it processes all the parking spaces in a particular frame in approximately 2 s. The system will also predict the peak time of the traffic so that users can check the availability of parking slots as per their convenience and book them in advance accordingly. Since the whole system is automated and does not need any human interference, therefore, it required a smart surveillance system for security of the parking area, which covers 13 real-world anomalies such as Accident, Stealing, Fighting,

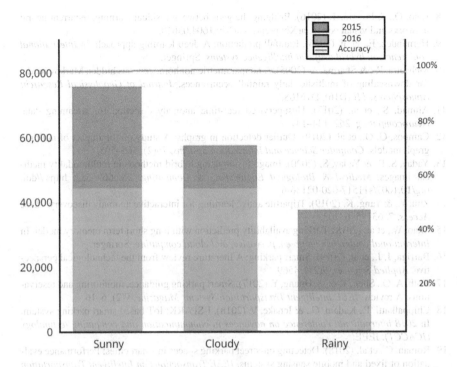

Fig. 5 Achieved accuracy corresponding to different weather images on which model is trained

Explosion, and many more. Thus, authors have successfully implemented a Smart Parking Management System.

References

1. Amato, G., et al. (2017). Deep learning for decentralized parking lot occupancy detection. *Expert Systems with Applications, 72*, 327–334.
2. Amato, G., et al. (2016). Car parking occupancy detection using smart camera networks and deep learning. In *2016 IEEE symposium on computers and communication (ISCC)*. IEEE.
3. Sultani, W., Chen, C., & Shah, M. (2018). Real-world anomaly detection in surveillance videos. In *Proceedings of the IEEE conference on computer vision and pattern recognition*. IEEE.
4. Yadav, S. P., Mahato, D. P., & Linh, N. T. D. (2020). *Distributed artificial intelligence: A modern approach* (1st ed.). CRC Press. https://doi.org/10.1201/9781003038467
5. Tran, D., et al. (2015). Learning spatiotemporal features with 3d convolutional networks. In *Proceedings of the IEEE international conference on computer vision*. IEEE.
6. Hou, R., Chen, C., & Shah, M. (2017). Tube convolutional neural network (T-CNN) for action detection in videos. In *Proceedings of the IEEE international conference on computer vision*. IEEE.
7. Yadav, S. P. (2020). Vision-based detection, tracking and classification of vehicles. *IEIE Transactions on Smart Processing and Computing, 9*(6), 427–434. https://doi.org/10.5573/IEIESPC.2020.9.6.427. SCOPUS, ISSN: 2287-5255.

8. Liao, Q., & Poggio, T. (2016). Bridging the gaps between residual learning, recurrent neural networks and visual cortex. arXiv preprint arXiv:1604.03640.
9. Hernández, E., et al. (2016). Rainfall prediction: A deep learning approach. In *International conference on hybrid artificial intelligence systems*. Springer.
10. Mehrotra, R., & Sharma, A. (2005). A nonparametric nonhomogeneous hidden Markov model for downscaling of multisite daily rainfall occurrences. *Journal of Geophysical Research: Atmospheres, 110*(D16), D16108.
11. Ahmad, S., et al. (2017). Unsupervised real-time anomaly detection for streaming data. *Neurocomputing, 262,* 134–147.
12. Campos, G. O., et al. (2019). Outlier detection in graphs: A study on the impact of multiple graph models. *Computer Science and Information Systems, 16*(2), 565–595.
13. Yadav, S. P., & Yadav, S. (2020). Image fusion using hybrid methods in multimodality medical images. *Medical & Biological Engineering & Computing, 58,* 669–687. https://doi.org/10.1007/s11517-020-02136-6
14. Zhu, Y., & Yang, K. (2019). Tripartite active learning for interactive anomaly discovery. *IEEE Access, 7,* 63195–63203.
15. Shao, W., et al. (2018). Parking availability prediction with long short term memory model. In *International conference on green, pervasive, and cloud computing*. Springer.
16. Barriga, J. J., et al. (2019). Smart parking: A literature review from the technological perspective. *Applied Sciences, 9*(21), 4569.
17. Kotb, A. O., Shen, Y.-c., & Huang, Y. (2017). Smart parking guidance, monitoring and reservations: A review. *IEEE Intelligent Transportation Systems Magazine, 9*(2), 6–16.
18. Chippalkatti, P., Kadam, G., & Ichake, V. (2018). I-SPARK: IoT based smart parking system. In *2018 international conference on advances in communication and computing technology (ICACCT)*. IEEE.
19. Roman, C., et al. (2018). Detecting on-street parking spaces in smart cities: Performance evaluation of fixed and mobile sensing systems. *IEEE Transactions on Intelligent Transportation Systems, 19*(7), 2234–2245.
20. Yadav, S. P., & Yadav, S. (2020). Fusion of medical images in wavelet domain: A hybrid implementation. *Computer Modeling in Engineering & Sciences, 122*(1), 303–321. https://doi.org/10.32604/cmes.2020.08459
21. Ramaswamy, P. (2016). IoT smart parking system for reducing greenhouse gas emission. In *2016 international conference on recent trends in information technology (ICRTIT)*. IEEE.
22. Yang, C.-F., et al. (2017). iParking–a real-time parking space monitoring and guiding system. *Vehicular Communications, 9,* 301–305.
23. Ma, S., et al. (2017). Research on automatic parking systems based on parking scene recognition. *IEEE Access, 5,* 21901–21917.
24. Jung, H. G. (2014). Semi-automatic parking slot marking recognition for intelligent parking assist systems. *The Journal of Engineering, 2014*(1), 8–15.
25. Shin, J.-H., & Jun, H.-B. (2014). A study on smart parking guidance algorithm. *Transportation Research Part C: Emerging Technologies, 44,* 299–317.
26. Mair, R. (2015). *How will city infrastructure and sensors be made smart*. The Cambridge Centre for Smart Infrastructure and Construction.
27. Bagula, A., Castelli, L., & Zennaro, M. (2015). On the design of smart parking networks in the smart cities: An optimal sensor placement model. *Sensors, 15*(7), 15443–15467.
28. Lin, T., Rivano, H., & Le Mouël, F. (2017). A survey of smart parking solutions. *IEEE Transactions on Intelligent Transportation Systems, 18*(12), 3229–3253.
29. Ngabo, C. I., & El Beqqali, O. (2016). Real-time lighting poles monitoring by using wireless sensor networks applied to the smart cities. In *Proceedings of the international conference on big data and advanced wireless technologies*. The Association for Computing Machinery.
30. Geng, Y., & Cassandras, C. G. (2012). A new "smart parking" system infrastructure and implementation. *Procedia-Social and Behavioral Sciences, 54,* 1278–1287.

31. Barone, R. E., et al. (2013). Architecture for parking management in smart cities. *IET Intelligent Transport Systems, 8*(5), 445–452.
32. Peng, G. C. A., Nunes, M. B., & Zheng, L. (2017). Impacts of low citizen awareness and usage in smart city services: The case of London's smart parking system. *Information Systems and e-Business Management, 15*(4), 845–876.
33. Hamidi, S. R., et al. (2017). Industry 4.0 urban mobility: goNpark smart parking tracking module. In *Proceedings of the 3rd international conference on communication and information processing*. The Association for Computing Machinery.
34. Mainetti, L., et al. (2014). Integration of RFID and WSN technologies in a smart parking system. In *2014 22nd international conference on software, telecommunications and computer networks (SoftCOM)*. IEEE.
35. Atif, Y., Ding, J., & Jeusfeld, M. A. (2016). Internet of things approach to cloud-based smart car parking. *Procedia Computer Science, 98*, 193–198.
36. Taherkhani, M. A., et al. (2016). Blueparking: An IoT based parking reservation service for smart cities. In *Proceedings of the second international conference on IoT in urban space* (pp. 86–88). The Association for Computing Machinery.
37. Mathew, S. S., et al. (2014). Building sustainable parking lots with the Web of Things. *Personal and Ubiquitous Computing, 18*(4), 895–907.

Credit Card Fraud Detection Techniques Under IoT Environment: A Survey

M. Kanchana, R. Naresh, N. Deepa, P. Pandiaraja, and Thompson Stephan

1   Introduction

Fraud in credit cards [36] is highly risky and a significant reason for monetary losses for firms as a whole as well as individuals. Due to the entire system being computerized, the data of credit cards and such transactions is colossally huge. Such huge data which is also skewed is very difficult to handle. Without data mining, data analysis, machine learning, and artificial intelligence techniques [37], it would have been a near-impossible task to detect such frauds manually. Before moving to ways to detect frauds, we need to understand the kind of fraudulent transactions that can occur. Although the frequency of fraudulent transactions is quite low compared to legal transactions, yet there are quite a few ways observed in which these frauds can occur. Fraudsters have a very dynamic nature and keep coming up with newer ways of performing frauds. A continued advancement is seen in strategies to commit frauds. It is thus of utmost importance to understand these techniques and detect frauds. Advances in various wireless networking protocols in technology such as 5G, RFID, Wi-Fi-Direct, Li-Fi, LTE, and other recent technologies [35] have

M. Kanchana · R. Naresh
Department of Computer Science and Engineering, SRM Institute of Science and Technology,
Kattankulathur, Chengalpattu, Chennai, Tamil Nadu, India
e-mail: kanchanm@srmist.edu.in; nareshr@srmist.edu.in

N. Deepa · P. Pandiaraja (✉)
Department of Computer Science and Engineering, M.Kumarasamy College of Engineering,
Karur, Tamil Nadu, India
e-mail: deepan.cse@mkce.ac.in; pandiarajap.cse@mkce.ac.in

T. Stephan
Department of Computer Science and Engineering, Faculty of Engineering and Technology,
M. S. Ramaiah University of Applied Sciences, Bangalore, Karnataka, India
e-mail: thompson.cs.et@msruas.ac.in

recently significantly improved the future functionality of IoT and made it more prevalent than ever, accelerating the further convergence of IoT with new technologies in other fields such as sensing, wireless recharging and data sharing. In addition, the privacy and security considerations that are present must be thoroughly investigated and discussed.

A deep understanding of such methods would help us make better and more effective and efficient systems to detect fraud. Certain ways in which frauds occur include:

(i) Theft of the card/physical loss of the card: Through such lost and stolen credit cards, maximum fraud happens. The CVV of the card is misused to make transactions. The owner might realize the theft late or may be delayed in alerting the bank for the same. This is quite a trivial way of committing scams.

(ii) Account takeover/application frauds: This kind of fraud is on the rise these days. Due to users' laxed security of online transactions, the scammers can get hold of users' sensitive information of the account. It is like identity theft. The fraudster can even apply for a new credit card [38] and/or open a new account in the name of the authorized personnel, in addition to reporting the actual card of the user as lost and blocking the same, thus gaining entire control over the account [2].

(iii) Phishing: As it is, phishing is a very severe cybercrime, and fraudsters can use it to get sensitive information without the knowledge of the user. Scammers bait their target with legitimate lookalike emails to unsuspected people. The only aim is to look quite legal and lure the people into interacting with them via means of shopping or anything else, generally by offering lottery or vouchers, etc. When a person believes or gets baited and passes their bank account information, they are easily misused.

(iv) Fraud applications/fraud merchant sites: Quite similar to phishing attacks, these fake sites and applications are a lookalike of genuine ones. The user's information is collected and later misused without the knowledge of the user.

(v) Mail interception/mail non-receipt fraud: The robber somehow gets access to card numbers by interfering somewhere in between the delivery. They can even get the card in their name to make purchases.

(vi) Card not present/card ID theft: It is a type of email or phone call fraud. It is quite reduced these days due to awareness of the theft. Yet, there are a few cases, where the scammers get card numbers, CVV and other sensitive information as per requirements to make transactions. Thus, the card identity is lost and misused by the fraudsters. They can easily make transactions without having the physical card with them.

(vii) Data breaching/collusion by merchant: A deliberate attempt by the merchant whom the user interacts with shares/leaks sensitive information of the customers. We are not habituated to question the card data storage before swiping. Awareness of the same is required, and developing such habits is necessary before swiping for transactions.

(viii) EMV technology/fraud cards: It is a way of forgery that still exists. Failure of upgrading the equipment reading the electromagnetic chips leads to this fraud. Scammers create fake cards, and older equipment can be fooled with the fake hologram. Skimming sensitive information on magnetic strips to steal sensitive information happens in this kind of fraud [2].

(ix) Altered/counterfeited cards: This is a very serious crime, when scammers create a fake magnetic strip using neodymium and hold original details of a user on it. These details are generally obtained from corrupt merchants. Such counterfeit cards are easily used to commit frauds.

(x) Skimming: This is a tedious task, yet can be very draining financially. Users' information is stolen by keeping small cameras or devices that can copy credit card information when swiping. For example, small cameras near ATM machines capture the pin entered by users, and the information is passed on to these scammers.

An accurate and efficient fraud system will have quite low misclassified data instances and correctly classify fraud transactions. It shall also have least or no legal transactions classified as fraudulent. A lot of research has taken place in fields of imbalanced data handling techniques. Classification techniques used include Neural Networks [4], Hidden Markov Models [5], Random Forests (RFs) and Support Vector Machines (SVMs) [6]. SVM and RF are considered to give better results compared to other algorithms, as these do not have mathematical constraints and statistical implications, rather they consider the previously misclassified instances and improve the current classification. SVM is based on statistical techniques but without implications and turns out to be quite successful in a wide range of problems [7]. Many ensemble techniques have also evolved through over the time, combining classifiers and/or bagging and boosting algorithms. Random Forests are seen to give better results compared to decision trees [8, 9] and quite equivalent results to SVMs [10]. Artificial Immune Systems (AIS) are used for data analysis and even as a classification algorithm. This is quite a new and recent branch of AI based on biological metaphors. It quite accurately distinguishes between its own kind and different instances. AIS predicts failing of corporate [11].

Class imbalance problems are recognized to be a huge issue while classification. Sampling being one of the most frequently used techniques to handle this problem, cost-sensitive methods are also gaining ground. Along with Neural Networks, moving the threshold of cost of cheaper or inexpensive class instances was proposed in [12]. Bagging and boosting of algorithms can also be performed to overcome this non-uniformity in data [13].

Python, a very popular language, is high level and has several libraries dedicated solely to image processing and face recognition. A number of techniques are also available for better recognition such as Karhunen-Loeve, Linear Discriminant Analysis (LDA) or Principal Component Analysis (PCA) [14]. SVM can also be used for recognition purposes. Feature positioning and extraction of the same can be done for selected areas. Also, to improve accuracy and efficiency, a majority voting of classifier results can be performed to give the final output [15]. A recent library addition in Python is done solely for face detection and recognition; this library is called face recognition library and has a backend of several classification techniques. We can choose any of these as per our requirements.

2 Datasets

2.1 US-Based Dataset

The dataset taken from [1] contains credit card fraud labelled data. This data is highly imbalanced with the majority of legal transactions. It has transaction description with nine variables, namely: custID, gender, state, cardholder, balance, numTrans, numIntlTrans, creditLine and fraudRisk.

2.2 German Bank Dataset

It consists of 21 columns:

- Instalment as income perc: Instalment to income ratio
- Personal status sex: marital status
- Other debtors: whether they are having any other debtors
- Property: what kind of property he/she is having
- Age: age of the person
- Other instalments: has the person made any other instalments
- Housing: whether they have their own house or not
- Credits the bank: credit line
- Present Res Since: time he is staying in this residence.
- Credits this bank: credit owed
- Job: person's job
- People under maintenance: number of people that are under maintenance
- Telephone: whether the telephone is registered or not
- Foreign worker: whether the person is a foreign worker or not

2.3 NUAA Database

NUAA (Nanjing University of Aeronautics and Astronautics) database contains collection of photographs taken from cheap web cameras [16]. It contains real and fake samples of 15 subjects. The database is created in three phases with two weeks of time in each phase. Photographs are taken in various illumination conditions. The database contains 2383 pictures of real faces and 3912 pictures of fake samples each having definition of 640 × 480 pixels.

3 Classification Algorithms

Quite a few classifiers can be used to classify the data into legal and fraud transaction. Class imbalance in almost all data of credit cards increases the problems of classification. Yet, these classifiers are the base for solving class imbalance problem. Generally, a class imbalance or class non-uniformity is handled using classifiers that deal with minority or outnumbered groups.

3.1 Support Vector Machine

An SVM Classifier's basic goal is to find hyper plane in n-dimensional data to accurately classify data instances. In this process, quite a few planes are selected. These work as decision boundaries to classify whether transaction lies on fraud or legal side. We prefer the maximized distance of data from decision boundary for better classifications [1, 3]. SVM works in high-dimensionality workspace without additional computational complexity. The way of classifying in multi-variable data is equally as simple as compared to linear classification by SVM. Non-uniformity of data is also handled by SVM quite well. Kernel representation by SVM for different kinds of data, and the [17, 18] property of maximizing the geometric margin for decision boundary by learning from various separating hyper planes make SVM quite beneficial to use.

3.2 Artificial Neural Networks

It is observed that neural networks perform better than quite a few other algorithms [19]. It has several neurons in each node. An input layer and quite a few hidden layers and one output layer, certain weights, chosen randomly are given to hidden layers depending bias. The summation of output from a particular layer is added to an activation function. This process is continued for each hidden layer. A final summation is carried out to give the result. The entire process depends on the kind of dataset and parameters and is adjusted automatically. With the help of back propagation, the weights first attached randomly are redistributed depending on outputs for the best possible results and improved efficiency.

3.3 Artificial Immune System (AIS)

It works like a classifier by training itself on a number of training examples. It has biological metaphors. For example, the training data is called antigen population, the process of detecting and classifying is called cloning and mutation [1]. There are about four extensions of AIS; each performs a very specific purpose. Artificial Immune Recognition System (AIRS) is very specifically used for classification purposes. Other extensions include Clonal Selection Algorithm (CLOALG), Immunos and Simple Artificial Immune System (SAIS). These others can also perform classification along with other functionalities. It randomly generates transactions that can be used to detect the distance between various instances. The distance from the detector assigns the class to the instances. These keep getting modified with every iteration of the algorithm form or reliable and better results. The appropriate distance that will act as threshold for classification has to be specified beforehand [19, 20]. Even the number of iterations for the algorithm needs to pre-defined. Euclidean distance formula is generally used for calculation purposes. The clonal algorithm works slightly differently, as it does not perform self-sampling, rather non-self or different set of instances is used for generating detectors, first randomly, then the distance used. The instances most distant from detectors are removed and added in place of random generation of detectors. This process repeats until detector set capacity is met.

3.4 Fuzzy Logic

In real world, we encounter situations where we cannot determine whether a particular state is false or true, and their fuzzy logic denotes a valuable flexibility for logic. By this process, we are able to consider the uncertainties and inaccuracies for any situation. Fuzzy term refers to things that are vague and unclear. The architecture [2] of the fuzzy logic consists of four parts:

- *Fuzzification*: It converts a set of inputs, that is, crisp numbers into a set of fuzzy sets. The crisp inputs comprises the proper inputs measured by using sensors, and after that, these values are passed to the control system be processed, like pressure, temperature, rpm, etc.
- *Rule Base*: Generally, it comprises a set of rules and the IF-THEN conditions provided by people to make a decision-making system based on the linguistic information. Modern developments are made in the fuzzy theory which provides several highly powerful methods for designing and tuning for fuzzy controllers. Mainly, these developments are used in order to reduce the frequency of fuzzy rules.
- *Inference Engine*: It decides the degree of matching the present fuzzy input w.r.t every rule, and after that, the decision is made about which rules are to be fired

based on the input field. After that, the fired rules are added in order to make the control actions.

- *Defuzzification*: It is basically used to translate the fuzzy sets that are acquired by inference engine and converted into crisp value.

4   Class Imbalance

Imbalance problem is a major issue, especially in case of credit card fraud datasets. The number of class instances for legal transactions is generally too high compared to fraud class instances. If this non-uniformity is not solved, classifiers will not be able to give proper, reliable results due to imbalanced training of the algorithm. In such cases, the classifiers will not be able to recognize fraud or scam and keep passing such scammers as legal or authorized users or transaction. Basically, the problem arising due to non-uniformity can be of two types: either to reduce false positives or when data collected is adequate for outnumbered class. Certain methodologies are mentioned in that it can be used to deal with an imbalanced data.

4.1   Boosting-Based Algorithms

These algorithms are weighted ones, that is, the algorithms in which the misclassified data of first iteration is weighted higher in the second iteration. They are a combination of a number of classifiers. Depending on all algorithms' individual performances, they are assigned the weights and importance levels, depending on dataset and parameters. This process of weighing the misclassified data higher in every following loop is called boosting of algorithm. These algorithms are either iterated for a preset number of times or keep running until a specified accuracy level is reached [13, 34]. Thus, this activity of weighing the wrongly classified data makes these algorithms quite independent of the imbalance in data. Since, at each iteration the weights are moved to erroneous or unusual behaviour, the minority class gets higher weightage, and thus, the algorithm does not show much variation pre- and post-upsampling or downsampling. These are quite strong classifiers that can be directly used on non-uniform data. Adaptive boosting is an example of boosting algorithm [23].

4.2   Bagging-Based Algorithms

Algorithms that help in variance reduction in every iteration are bagging-based. It does so by increasing the number of training samples by using combination of repeated instances. Partitioning of data is done quite randomly with the sole goal of reducing variance. Average of all predictions is taken as the final classification. Simultaneously, a number of data instances are replicated and divided into groups. These groups are used to train the classifier. The result is an ensemble model which is more robust than the single classifier. Also, such a model solves the problem of overfitting and can capably handle the high dimensionalities of data with high accuracy even for missing and erroneous data entries [13, 24, 34].

4.3 Changing Class Distributions

Re-sampling methods are used to alter the structure of a training dataset, especially for an imbalanced classification task. Main attentions for re-sampling methods on imbalanced classification are used for oversampling of the minority class. Also, a set of techniques is developed specially for undersampling of majority class, which is also used with effectual oversampling methodologies.

- *Undersampling*: Undersampling [13] is the process of removing samples from the majority class from the training dataset at random. In the modified version of the training dataset, this reduces the number of cases in the majority class. This procedure can be continued until the desired class distribution, such as an equal number of instances for each class, is obtained.
- *Oversampling Techniques:* Random oversampling [24, 25, 34] entails replicating minority class instances at random and adding them to the training dataset. Replacement is used to choose examples from the training dataset at random.

4.4 Cost-Sensitive Learning (CSL)

This type of machine learning takes the misclassified costs into consideration. The main aim of this learning is to reduce the whole cost. The main difference between the cost-insensitive learning and cost-sensitive learning is based on how they treat the different misclassifications differently. It takes misclassified costs into consideration. In cost insensitive learning the examples into a set of known classes. In many real-world applications, class imbalance problems occur in which the distribution of data that is present in the dataset is highly imbalanced. It has the structure of a cost matrix and relates to a sample from a true class I to a matrix entry. The matrix is expressed in terms of the problem's average misclassified costs. The elements in the diagonal of the matrix are set to zero, indicating that proper categorization comes at no cost. The primary goal of cost-sensitive classification is to lower the cost of mis-classifications, which may be accomplished by selecting the class with the lowest conditional risk [3, 13, 32].

4.5 One-Class Learning (OCC)

It is based on a recognition method, which is an alternative to discrimination in which the model is exclusively dependent on the target class instances. In this case, the classification is done by adding a entrance on the similar value [26] between the target class and query object. These two classes of learning have been considered previously in the conditions of identification-based one-class approach: SVMs [26, 28], auto-encoders [2, 28]; they were competitive as observed in [27, 28]. Systems that learn only the minority class may still be trained by the examples that belong to every class. Ripper [31], Shrink [30] and Brute [29] are three machine learning systems that are similar. Brute was used to check for errors in the process of identifying credit card fraud [29]. Shrink [30] takes a similar method. Shrink assigns positive classifications to mixed regions (i.e. regions containing both negative and positive examples) based on the expectations that there will be more genuine transactions than fraudulent. Ripper [31] is a rule induction system that repeatedly creates laws to complete previously unfinished training instances [33] using a separate-and-conquer strategy. Every legislation is formed up of conditions that are added until no legal transactions are affected. It generally establishes regulations for each class, starting with the most unusual and ending with the most common. As a result, Ripper may also be considered a one-class learner in this perspective. One interesting feature of one-class learning is that it may solve the classification issue faster than discriminative (two-class) methods in particular cases, such as multimodality of domain space [26].

5 Face Recognition Techniques

There are a number of ways of performing face verification. The basic concern is to detect whether it is a live image or a fake one. For this purpose, the given methodology is followed (Fig. 1).

5.1 Dynamic Approach of Extracting Features

This kind of technique is first among itself to use a two-dimensional planar method to detect fake. It works upon studying the features of face-like movement of eye balls and looking for the center point for the eyes and then for the entire face after

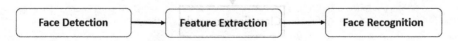

Fig. 1 Basic face recognition method representation

normalizing the eyes. Various features of face are extracted, although, the prime focus for this algorithm are the eyes for detecting the liveliness. It focuses on eyes and compares the eyes in dataset to detect real or fake image. Form a king the eyes look as real as possible in a three- dimensional view a number of techniques, called filtering can be applied. It is seen Gaussian filtering performs quite well in this extraction. Ensemble and/or boosting algorithms can be used to as classifiers. Figure 2 [15] represents the process.

5.2 Colour Base Extraction

This extraction of face and then recognition can be done in two ways. Either the skin colour filter is applied to extract the area or background colour is used to extract all the remaining portion of the image. In both the cases, the risk of image damaging and lower quality of results is higher as some parts of face that are too bright or dark might not get extracted. For example, the area under around eyes is comparatively darker and thus might not get extracted. Also, variable lighting can be a challenge to

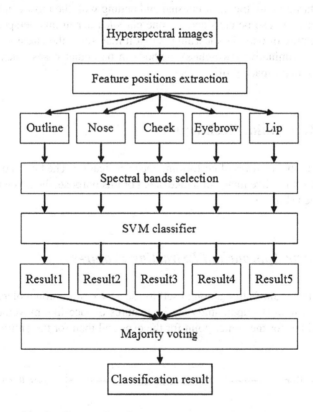

Fig. 2 Process of the face liveness detection

handle. Healey [21] proposed that the input image should be in RGB format with 0–255 values for caricature to deal with this scenario. RGB matrices are then made to act like a wet blanket desaturation, when image is being changed from RGB to IRgBy to signify space. In case of very complicated backgrounds, MUHULANOBIC metrics is used [22].

5.3   Cam Shift Algorithm

This is also called Continuously Adaptive Mean (CAM) Shift Algorithm. It helps reduce the complications of reducing image noise. It simplifies usage of black box for detecting objects. It starts after setting a calculation region at search window but in a greater size. Then, the mean shift is executed for a given number of times. It calculates the mean location in search window and centres the window at mean location calculated in previous iteration. Until converging, the iterations are repeated. Finally, the window for search is set up at a size as in formulas (1) and (2):

$$M00 = \Sigma\Sigma(x, y) \tag{1}$$

$$x\, y$$

$$S = 2 * \mathrm{sqrt}(M00 / 256) \tag{2}$$

Here, $M00$ represents zeroth moment for the 2D image of dimensions x and y, $I(x,y)$. S represents the size of the window for search. Although this algorithm has been improved over the years, better results can be obtained using deep learning methods.

5.4   Convolutional Neural Networks

A widely used neural network based on deep learning is used nowadays for face recognition. Performance, reliability and efficiency are quite high in this one. A hierarchy of layers is made for hidden layers, and reduces the pre-work on input before entering the network. For training purposes, a huge number of images are cropped/trimmed and resized. Then these images are stuck together in face dataset where each part has two people. Gray degree treatment is given to dataset. Several models can be made for training."LeNetConvPoolLayer" is one of them. In this, the third hidden layer has a sampling layer pre-fitted. A simple Softmax classifier is used for classification. Figure 3 represents a basic CNN structure for the model.

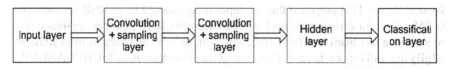

Fig. 3 CNN structure

5.5 Face Recognition Library

A specialized library in Python for face detection and recognition has been recently developed. It is built upon Dlib, a toolkit with ML algorithms for solving real-time complex problems; it is written in C++. It is a very simplified way of performing face recognition using most complicated algorithms at backend. For example, if CNN is required to perform the image recognition, it can simply be called in the single line of code. Tolerance and sensitivity for the amount of strictness required in the algorithmic functions can also be set manually. A GPU enhances the performance of the library manifold. This has a very high accuracy of around 99%.

6 Conclusion

The classification and data handling methods are quite secure in terms of loss of data. Depending on data to data, these mechanisms can be utilized for best possible results. Face recognition techniques are used for secured and sheltered outputs. Also, rightful validation and fraud detection can be improved using these techniques. AIS are still under development. Face recognition library is also a very new module in Python, which with time will be improved for better results. These are ever-evolving fields. Yet, absence of datasets with both credit card details and images of bank account holders is a major challenge yet to be conquered. Grave concerns regarding fraud detection are present globally. It is a very critical issue at hand to keep up with dynamicity of fraudsters and their ways of committing these scams. This chapter has summarized certain classification algorithms, followed by ways of handling class imbalance problem and various techniques of face detection and recognition that can be extended for the face verification module.

References

1. Makki, S., & Assaghir, Z. (2019). An experimental study with imbalanced classification approaches for credit card fraud detection. *IEEE Access, 7*, 93010–93022.
2. Dubey, S., Jain, Y., Jain, S., & Tiwari, N. (2019). A comparative analysis of various credit card fraud detection techniques. *International Journal of Recent Technology and Engineering, 7*, 402–407.

3. Tharakunnel, K., Jha, S., Siddhartha, B., & Westland, J. C. (2011). Data mining for credit card fraud: A comparative study. *Decision Support Systems, 50*, 602–613. Elsevier.
4. Rao, B., Freisleben, B., & Aleskerov, E. (1998). A neural network-based database mining system for credit card fraud detection. In *Proceedings of the computational intelligence for financial engineering (IAFE)*. IEEE.
5. Majumdar, A., Kundu, A., Srivastava, A., & Sural, S. (2008). Credit card fraud detection using hidden Markov model. *IEEE Transactions on Dependable and Secure Computing, 5*, 37–48.
6. Hand, D. J., Whitrow, C., Juszczak, P., Weston, D., & Adams, N. M. (2009). Transaction aggregation as a strategy for credit card fraud detection. *Data Mining and Knowledge Discovery, 18*, 30–55.
7. Shawe-Taylor, J., & Cristianini, N. (2000). *An introduction to support vector machines and other kernel-based learning methods*. Cambridge University Press.
8. Kuncheva, L. I. (2005). *Combining pattern classifiers with inter science methods and algorithms*. Wiley.
9. Polikar, R. (2006). Ensemble based systems in decision making. *IEEE Circuits and Systems Magazine, 6*, 21–45.
10. Niculescu-Mizil, A., & Caruana, R. (2006). An Empirical Comparison of Supervised Learning Algorithms. In *Proceedings of the 23rd international conference on machine learning*, Pittsburgh, Pennsylvania.
11. Brabazon, A., et al. (2006). Financial classification using an artificial immune system. In K. E. Voges & N. K. Pope (Eds.), *Business applications and computational intelligence* (pp. 389–406). Idea Group Publishing.
12. Zhou, Z., & Liu, X. Y. (2006). Training cost-sensitive neural networks with methods addressing the class imbalance problem. *IEEE Transactions on Knowledge Data Engineering, 18*, 63–77.
13. Feng, W., Huang, W., & Ren, J. (2018). Class imbalance ensemble learning based on the margin theory. *Applied Sciences, 8*, 815.
14. Paunikar, S., Sheikh, A., Admane, A., Jawade, S., Sawarkar, M. J., & Wadbude, S. (2019). A review on different face recognition techniques. *International Journal of Scientific Research in Computer Science, Engineering and Information Technology, 5*(1), 207–213.
15. Liu, Y., Zheng, M., & Li, Q. (2017). Face liveness verification based on hyper-spectrum analysis. In *2017 31st international conference on advanced information networking and applications workshops (WAINA)*. IEEE.
16. Yan, Q., Li, Y., Deng, R. H., Xu, K., & Li, Y. (2014). Understanding OSN-based facial disclosure against face authentication systems. In *Proceedings of ACM Asia symposium on information, computer and communication security (ASIACCS)*. ACM.
17. Pise, N., & Kulkarni, P. (2016). Algorithm selection for classification problems. In *SAI computing conference*. IEEE.
18. Le Borgne, Y.-A., Pozzolo, A. D., Waterschoot, S., Bontempi, G., & Caelen, O. (2014). Learned lessons in credit card fraud detection from a practitioner perspective. *Expert Systems with Applications Journal, 10*, 4915–4928. Elsevier.
19. Bhaduri A. (2009). Credit scoring using Artificial Immune System algorithms: A comparative study. World Congress on Nature & Biologically Inspired Computing (NaBIC), 1540–1543.
20. Brabazon, A., Cahill, J., Keenan, P., & Walsh, D. (2010). Identifying online credit card fraud using Artificial Immune Systems. IEEE Congress on Evolutionary Computation, 1–7.
21. Robertson, D. J. (2015). Face averages enhance user recognition for smartphone security. *PLoS One, 10*, e0119460.
22. Kaehler, A., & Bradski, G. (2008). *Learning OpenCV computer vision with the OpenCV library*. O'Reilly Media, Inc.
23. Wang, J., & Li, Z. (2018). Research on face recognition based on CN. *IOP Conference Series: Earth and Environmental Science, 170*, 1–5.
24. Zhou, G., Yin, Y., Yang, G., Guo, X., & Dong, C. (2008). On the class imbalance problem. In *Fourth international conference on natural computation*. IEEE.

25. Chouhan, T., & Sahu, R. K. (2011). Classification technique for the credit card fraud detection. *IJLTET, 10*, 283–286.
26. Japkowicz, N. (2001). Supervised versus unsupervised binary learning by feed forward neural networks. *Machine Learning, 42*, 97–122.
27. Williamson, R. C., Shawe-Taylor, J., Scholkopf, B., Smola, A. J., & Platt, J. C. (2001). Estimating the support of a high dimensional distribution. *Neural Computation, 13*, 1443–1472.
28. Tax, D. (2001). *One-class classification*. Ph.D. dissertation, Delft University of Technology.
29. Yousef, M., & Manevitz, L. M. (2001). One-class SVMs for document classification. *Journal of Machine Learning Research, 2*, 139–154.
30. Etzioni, O., Segal, R., & Riddle, P. (1994). Representation design and brute-force induction in a Boeing manufacturing design. *Applied Artificial Intelligence, 8*, 125–147.
31. Matwin, S., Kubat, M., & Holte, R. (1997). Learning when negative examples abound. In *Proceedings of the ninth European conference on machine learning* (LNAI) (Vol. 1224, pp. 146–153). Springer.
32. Cohen, W. W. (1995). Fast effective rule induction. In *Proceedings of the twelfth international conference on machine learning* (pp. 115–123). Elsevier.
33. Tomek, I. (1976). Two modifications of CNN. *IEEE Transactions on Systems Man and Communications, 6*, 769–772.
34. Prasad, A., & Rustogi, R. (2019). Swift imbalance data classification using SMOTE and extreme learning machine. In *Second international conference on computational intelligence in data science*. IEEE.
35. Choi, D., & Lee, K. (2018). An artificial intelligence approach to financial fraud detection under IoT environment: A survey and implementation. *Security and Communication Networks, 2018*, Article ID 5483472, 15 pages. https://doi.org/10.1155/2018/5483472
36. Rai, A. K., & Dwivedi, R. K. (2020). Fraud detection in credit card data using machine learning techniques. In A. Bhattacharjee, S. Borgohain, B. Soni, G. Verma, & X. Z. Gao (Eds.), *Machine learning, image processing, network security and data sciences. MIND 2020* (Communications in Computer and Information Science) (Vol. 1241). Springer. https://doi.org/10.1007/978-981-15-6318-8_31
37. Priscilla, C. V., & Prabha, D. P. (2020). Credit card fraud detection: A systematic review. In L. Jain, S. L. Peng, B. Alhadidi, & S. Pal (Eds.), *Intelligent computing paradigm and cutting-edge technologies. ICICCT 2019* (Learning and analytics in intelligent systems) (Vol. 9). Springer. https://doi.org/10.1007/978-3-030-38501-9_29
38. Kundu, A., Sural, S., & Majumdar, A. K. (2006). Two-stage credit card fraud detection using sequence alignment. In A. Bagchi & V. Atluri (Eds.), *Information systems security. ICISS 2006* (Lecture notes in computer science) (Vol. 4332). Springer. https://doi.org/10.1007/11961635_18

Trustworthy Machine Learning for Cloud-Based Internet of Things (IoT)

Saumya Yadav, Rakesh Chandra Joshi, and Divakar Yadav

1 Introduction

Recent years have seen an explosion of research in technologies for trustworthy machine learning in the Internet of Things (IoT) that introduce requirements around privacy, human rights, fairness and other properties of AI-based services [1]. The applications of IoT-based services and products goes through every industry and sector from a smart city, mining, education, transportation, manufacturing, infrastructure management, smart home, commerce, health care, surveillance, utilities, to logistics and supply chain. The possibilities obtained by the IoT are countless, and it will be more challenging in the future where new devices are getting connected through the Internet day by day. Although the advantages of IoT are unquestionable, data security is still the matter of concern. IoT is so confronting because it permits the objects to be controlled or sensed remotely through an existing network. It creates endless possibilities to directly integrate with computer-based systems in the physical world. Once sensors, robots, drones, cars, and humans are adequate to effortless mutual interaction from anywhere around the world through IoT, infinite number of threats will be exposed. For example, researchers have found wide range of perilous vulnerabilities in IoT baby monitoring [2], whose advantage can be taken by the hackers to carry out iniquitous activities such as authorization to other users to remote access of the system and for viewing the monitor. When it comes to demanding and delivering assistance, IoT resources work together. Such services must be able to trust one another in heterogeneous and dynamic environments.

S. Yadav · R. C. Joshi (✉)
Centre for Advanced Studies, Dr. A.P.J. Abdul Kalam Technical University, Lucknow, India
e-mail: saumya@cas.res.in; rakesh@cas.res.in

D. Yadav
National Institute of Technology, Hamirpur, India
e-mail: divakaryadav@nith.ac.in

© The Author(s), under exclusive license to Springer Nature Switzerland AG 2022
F. Al-Turjman et al. (eds.), *Transforming Management with AI, Big-Data, and IoT*, https://doi.org/10.1007/978-3-030-86749-2_9

Behaviour-based attacks, as discussed by Bhardwaj et al. [3], pose a threat to IoT trust protection by causing nodes to spontaneously execute good and bad actions in order to prevent being classified as a threat. Other vulnerabilities are found in Internet-connected cars, which may be controlled remotely. It was proven in another work that IoT-based cars which were connected through the Internet can be controlled remotely, having different functions such as shutting down the car while in motion, unlocking the car doors [4]. Some other troublesome IoT hacks include hacker's involvement in medical devices, which may have lethal consequences on patients' health [5, 6]. Some important features for next-generation secure IoT systems are as follows.

Machine Learning Conventional methods for IoT security focus to address particular threats under the specific network conditions. Nevertheless, every threat cannot be addressed and also malicious activity can also be dynamic in its nature. Cross-layer attack [7, 8] is an example of one such attacks, which have objectives and activities involving different layers of network protocol stack. The attacker may choose different layers to attack rather than choosing a target layer, in a cross-layer attack. Thus, a small-scale attack may lead to intense deviations on the target layer. To tackle such threats, machine learning is an encouraging way to a development of such security systems which will learn towards the efficient detection and mitigation of dynamic attacks.

Security-by-Design There is a requirement to build a secure IoT device from scratch, where the device is free from all the vulnerabilities, with different measures like performing continuous testing, adherence of best practices, and safe authentication. It is done to tackle existing threats and vulnerabilities as well as to patch security holes because these things are comprised of hit-and-miss process and are required to be effective to secure IoT from scratch. Building a secure IoT network at the designing phase lessens some potential disruptions and escapes from an expensive and difficult attempt to provide the security to different products after their development.

Polymorphic Securing the IoT for the next generation must be integrated in both application software layers and in hardware. Moreover, bolted-on security mechanisms have no potential to provide resistance to dynamic attacks in IoT in multiple levels of the network system. So, software-defined networking (SDN) can be utilized for creation of technical designs, and implementation of context-aware and polymorphic security measures that have ability to sense and respond to different kinds of attacks by making a change in software and hardware structures of IoT devices.

Recent advances in artificial intelligence and machine learning (ML) allow IoT to learn and accordingly support the preferences and lifestyle of an individual at work, at home and while moving. Small sensors will be able to capture biometric information with remote access [9] and give clinical assistance with detailed description about the health condition, and can effortlessly deliver medicines [10].

Mutual information will be exchanged between IoT and the person using machine learning. Contrastingly, sharing a personal information about own or any other business-related information may not be safe as it can be accessed by unwanted third parties or some malicious users. Hence, privacy and restriction of access became a significant aspect of IoT [11, 12]. Humans are playing the primary role to realise the sensing process in the IoT. But still, there is no such assurance that a consistent information will be generated by the humans, in this regard [13]. To solve this issue, reputation mechanisms and novel trust will be required to scale current human populations [14].

To build trustworthy ML systems for the cloud-based Internet of things, one needs to include trustworthiness [15] in all stages of the design, that is, from collecting data to selecting algorithm and visualizing results. To make the system work properly, ML algorithms need to have a consistent and clear formalization of the states (i.e. not attack/attack), inputs (i.e. image, sound data), and outputs (i.e. models, processed data). A clear explanation of the input is to apply the theory of dimensionality reduction, by performing the task of feature extraction and feature selection [16]. Feature extraction consists of procedural method to extract useful features from the set of available features. There are three different strategies that are mostly followed for feature extraction. The first one is the embedded strategy where feature selection is done for removal or an addition based on prediction error while developing a model. The second one is the wrapper strategy where feature search is done based on the accuracy parameter. The third is the filter strategy which took information gain as a feature selection criteria. Another major challenge for ML for cloud-based IoT is to know about the attack and type of attack. In other words, we can characterize and formalize as:

1. "Bad" IoT network state when attack is taking place
2. "Good" IoT network state when there is no attack

Efforts of this category have been shown in another work [17], where the Bayesian learning is controlled for creating a relation between the probability of occurrence of an attack and evidences supporting that attack. Moreover, if an attacker has access to collected training dataset, or knows the detail about the algorithms used, the attacker can tune the model while training and generate a malicious model. Mostly, it is equitable to consider that such data would not be available, because unsupervised machine learning algorithms are dependent on the parameters which could be encoded on the hardware chip of the IoT devices, and another reason is that the supervised ML algorithms would be trained before placement of the IoT nodes. Nevertheless, the effect of such kinds of attacks must be investigated in further research.

2 Related Work

ML is a power tool which is used in every field to solve the critical problem. IoT is one of the fields which uses ML to solve the security issues. Lately, Zhang et al. [17] have proposed a framework based on Bayesian learning application which mitigate and detect cross-layer wireless attacks. Precisely, it establishes a probabilistic relation between the supporting evidence such as the indication of malicious attack in the network and a hypothesis where the malicious attack is probably taking place. Whenever such evidences are found, the hypothesis are updated dynamically. Consequently, the more accurate is the resulting hypothesis on getting the more evidence. Through simulations and experiments, the author proved that even minor unwanted activities in the process can be detected with high confidence, when abundant of evidences are collected. A learning mechanism is proposed by Cañedo and Skjellum [18] to evaluate the strength of data generated by IoT devices using neural networks. Likewise, ARTIS is proposed by Hofmey and Forrest in [19], which is an artificial immune system which encourages ML algorithms to develop an adaptive immune system.

An intelligent system is developed in [20, 21] using ML to automatically detect any malicious activity in the network and other security threats. Vulnerability assessment mechanism is proposed by Miettinen et al. [22] that used ML to classify and identify IoT devices according to their trustworthiness.

3 Machine Learning and Internet of Things (IoT)

3.1 *Machine Learning*

Artificial intelligence field has gone through huge advancements in the recent decade. Machine learning [23], which is a branch of artificial intelligence, is defined as a study of computer programs that allows computers to understand from the experiences and examples, without being explicitly programmed. Multiple tasks can be achieved with the help of machine learning such as pattern recognition, predictions, clustering, classification, etc. Machine learning models are trained using various algorithms, and analysis is done in different parameters. Computational systems are trained through different algorithms and statistical models for analysis of the sample data, to develop an efficient learning process. Domain knowledge is required to decide features in machine learning algorithm can work properly for improving the predictions. Features are the variable or descriptive attributes that are recorded and quantified from raw data to train machine leaning model. The output values or predictions are called labels. Machine learning algorithms try to identify the patterns using the information obtained from the raw data and will be used for prediction of the new data. Machine learning algorithms are characterised into four different types: supervised learning, semi-supervised learning, unsupervised learning and reinforcement learning.

Supervised Learning Supervised learning is a kind of machine learning that tries to map an input with an output using a labelled training data and generates some function for prediction of unseen data. Supervised learning is employed for different machine learning-related problems including regression and classification. Various machine learning algorithms such as Support Vector Machines (SVM), Random Forest, Decision tree, Naïve Bayes and k-Nearest Neighbour classifiers are employed to perform different tasks like population growth prediction, weather forecasting, speech recognition, digit recognition, etc. The data used for training with the known labels and supervised classifier try to establish the relation between the input variables or features with data labels.

Unsupervised Learning Unsupervised learning is a kind of machine learning technique that tries to identify the undetected patterns in input dataset without providing any labelled response. Cluster analysis is the most common unsupervised learning method for experimental data analysis to discover hidden patterns or data organization. Principal component analysis and clustering are two important methods of unsupervised learning. Euclidean or probabilistic distance is used as a measuring criteria to find the similarity in data points which are modelled into clusters. Gaussian mixture models, self-organizing maps, k-means clustering and hidden Markov models are few of the important clustering-based unsupervised algorithms.

Semi-supervised Learning Semi-supervised learning is a unified form of both supervised and unsupervised learning methodologies. It is applicable to both labelled and unlabelled data. Labelled data is in large portion compared to unlabelled data. It substantially improvises the performance of the machine learning classifier. Graph-based methods, low-density separation, generative models, and heuristic approaches are used for the task of semi-supervised learning.

Reinforcement Learning Reinforcement learning mainly deals with software agents to take actions in a particular situation in order to maximize the notion of cumulative reward. Labelled input–output pairs are not needed in reinforcement learning, unlike supervised learning. In short, the output of the generated model depends upon the state of present input whereas the next input value relies on the output from the previous input. As the decision is dependent in reinforcement learning, labels are given to sequences of dependent decisions. Reinforcement learning has major applications in real-time decisions, robot navigation and AI gaming.

3.2 Internet of Things (IoT)

The Internet of things (IoT) characterizes the interconnection of physical objects or "things" which consists of sensors, processing networks and software to connect and exchange the data with other devices and systems over the Internet. Sensors and other connected devices collect the data from the given environment, and information is extracted from the raw data. Then, the information is transferred to cloud servers and other devices via the Internet. In a smart city, the IoT detects and disseminates critical data. Implementing energy efficiency in smart cities, as implemented by the authors in [24], necessitates securing IoT products, which is a difficult task given the available energy budget. Trust protection is a simple technology that can be used to secure IoT computers.

IoT is aimed at developing a smart environment and simplifies routine life by saving money, energy and time. Industries running on IoT reduced the expenses, and this field has seen huge growth in the last decade. IoT devices transfer the data to one another to enhance the performance of the system in an automatic manner without the interference of humans. RFID tags create an advancement and are used frequently nowadays, and low-cost sensors [25] have become more available, and protocols have been changed with new developed web technology. Functioning of the systems is further enhanced after integrating with different technologies with better connectivity among the devices. Thus, protocols need to be modified accordingly with the advancement in the technology. There are three type of communication protocols in terms of IoT, that is, device to device (D2D), device to server (D2S) and server to server (S2S). D2S communication protocols are used for the complete cloud processing-based tasks. S2S communication, which is applied mostly in cellular networks, is challenging in preparing and processing data. Fog and cloud processing are two of the main methods used to process and prepare the data before transferring to other devices. The architecture of cloud-based ML models for IoT is shown in Fig. 1.

4 Trustworthy Machine Learning for Cloud-Based IoT

Cloud computing gives advantages to IoT by enabling remote accessing its devices using Internet connection. The integration of cloud with IoT offers abundant opportunities for IoT-based systems. IoT can take advantage of enormous resources of the cloud like data sharing, making predictions, accessing devices, etc. using Internet connection. Cloud can use IoT device as a link which is integrated into real-life applications through a distributed means. It supplies cloud services to large IoT consumers by covering a larger area. IoT when utilizing cloud services becomes a distributed system, which is vulnerable to various unwanted attacks and introduces privacy concerns for the users. Unwanted attacks can manipulate the data security and get access to some confidential data. It manipulates the data while transferring

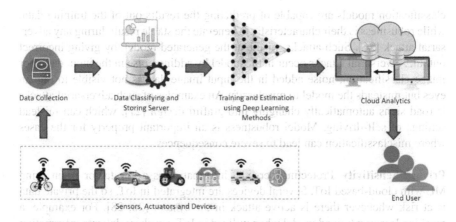

Data Collection   Data Classifying and   Training and Estimation   Cloud Analytics
                   Storing Server   using Deep Learning
                                           Methods

Sensors, Actuators and Devices   End User

Fig. 1 Architecture of cloud-based machine learning models for Internet of Things

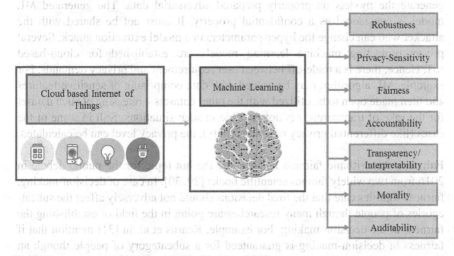

Fig. 2 Trustworthy machine learning for cloud-based IoT

it from different IoT devices, can bring insufficient integrity control on the data, and other security threats. Thus, we need a technology to ensure the security of data for cloud-based IoT. ML get advanced over the years and generate several smart applications on the basis of previous data. Since cloud-based IoT produces huge amount of data, it can be utilized by ML algorithms to generate models for predicting secure or nonsecure IoT devices. It is a smart technology which intelligently monitors IoT devices and secures it from unwanted attacks. There are certain labels for cloud-based IoT which ensure trustworthiness of machine learning as shown in Fig. 2.

Robustness Robustness is the ability of machine learning models to handle the errors during its execution for cloud-based IoT. Machine learning classification models are capable of predicting the result out of the training data. Machine learning classification models are capable of predicting the results out of the training data, while robustness is their characteristic to generate the stable result during any adversarial attack [26]. Such attacks misdirect the generated model, by giving incorrect output. Adversarial attacks occur in the model by adding noise in the input data. For image classification, noise added in the input image data is not visible to human eyes but misleads the model to misclassify. An examples of such adversarial attacks is road signs automatically changing into graffiti design [27], which can mislead settings of self-driving. Model robustness is an important property for the cases where misclassification can lead to severe consequences.

Privacy Sensitivity Protecting personal information is a key factor when using ML with cloud-based IoT. Several devices are integrated in IoT, so the private data is at risk whenever there is active attack in the system [26, 28]. For example, a machine learning trained model when utilized in IoT can share delicate information to an unknown attacker about the training dataset. These unknown attackers can generate the models on properly prepared adversarial data. The generated ML model is considered as a confidential property. It must not be shared with the attacker who can change the hyper-parameters in a model extraction attack. Several privacy-protecting machine learning models are established for cloud-based IoT. Hence, there is a trade-off between user requirement and privacy demands. For example, ML algorithm may be trained with data comprising of sensitive features and then made open to be utilized with the other datasets – releasing of such dataset for the sake of transparency is not possible in such situations. Still in some of the cases (like differential privacy or k-anonymity), the privacy level can be calculated.

Fairness Algorithmic fairness has become the hot topic for the public debate in 2016 from two widely famous scientific books [29, 30]. In case of decision-making, fairness signifies care that the final decisions should not adversely affect the subcategories of people though many researches are going in the field of establishing the fairness in the decision-making. For example, Kearns et al. in [31] mention that if fairness in decision-making is guaranteed for a subcategory of people though an individual attribute, fairness is not ensured when that particular attribute combines with other attribute. Liu et al. in [32] demonstrated 21 fairness criteria, and two criteria out of them pointed out that fairness criteria can cause some harm in longer haul.

Malicious algorithmic bias can be measured in quantitative manner, for example, true positive and false negative can be taken as estimation parameter to check the deviation from the ideal output [33]. Fairness criteria hold an important role to decide fairness, and there are chances of some delayed effect which cannot be computed in an immediate instant.

Thus it is requirement for selection of fairness criteria and to define the distinguished subgroups for transmission of fairness of ML model to the end user for which fairness criteria is to be estimated.

Accountability There are lots of decision-making models, and it is demanded each of the decision-making agents should be fully responsible for all their roles and decisions. Doshi-Velez et al. in [34] suggested three different tools for achieving the accountability for the algorithms.

The first one suggests that theoretical guarantee can be used for both problems and results to make them fully formalized. In the second one, statistical evidences are much more suitable in those situations where output can be measured and formalized, but there is nonavailability of sufficient prior knowledge of the complete issue. Third one is for those problems which are not completely described. In these cases, accountability can be handled through these explanations, and these kind of views are also taken by the European Parliament Research Service [35]. Legislation and its operationalization will be required to provide the accountability of machine learning models to end users. If both legislation and its functioning were made available, ML models utilized in cloud-based IoT could be made certified so that a certificate can be given to each customer label.

Transparency/Interpretability At present, model transparency is in major demand and is also recognised by the EU General Data Protection Regulation [36, 37]. Miller in [38] defines interpretability as "the degree to which a human can understand the cause of a decision". Few of the ML model classes are fundamentally transparent by default and are considered as interpretable [39], for example, decision rules [26]. Post hoc explanation strategies have been established for complex ML generated models [40]. Interpretability is dependent on an individual background knowledge, so it is inherently subjective. It is thus best estimated in real-world situations [41]. Since these estimations are not affordable, various proxies have been proposed. Decisions can be transferred to the end users through the recent research outcomes by means of artificial intelligence.

Morality Freeman et al. merged different moral concepts in their kidney donor exchange program for deciding the donor matches with the help of artificial intelligence [42]. The judgment of matching behaviour and morality differs in societies and within cultures [43]. Based on 40 million decisions from the "moral machine experiment", where users encountered various moral dilemmas in the milieu of a hypothetical self-driving vehicle, the above decision has been made [44]. In a same manner, morality of ML-based applications can be judged from the consumer labels for an end user with social acceptability and trust.

Auditability The ML auditability method enables the third party to verify the output and function of the generated model. The inspiration for auditing in ML came from three other properties: safety and security, transparency, and fairness. For execution of regulations such as European GDPR legislation, auditability of the model generation and function is important. GDPR is about protecting the data, posing requirements on government, industry and other data owners about safeguarding the transparency of the data collection process and unambiguous agreement from the person for the use of his data. Under GDPR process, the individual has right to challenge the data collection, have complete right on their collected data and uphold the "right to be forgotten" from the collected data source [45]. GDPR also requires a suitable explanation for any decision made by ML algorithm [46]. Mentioning the procedures to audit the model, and the process used to generate the model confirms that the model is accountable for giving the final output. Accountability in making decision by different models has also emerged outside GDPR [47] and is under review in other countries like India [48] and China [49].

5 Conclusion

IoT associates billions of smart devices that can interconnect with each other having a minimum human interference. It plays a vital role in enhancing numerous smart applications which can improve the living quality of a person. With these benefits, it also brings some privacy and security concerns like network security, encryption, application security, access control, integrity, and authentication of connected devices. The requirement of securing cloud-based IoT devices is a trending research topic as many transmission devices, home appliances, and other devices are connected through it. The development of ML technology enables to secure these devices by analysing the produced IoT data and generating prediction models for securing the IoT connections and devices. Cloud-based IoT produces huge data which can be used to generate ML models for classifying secure network and connecting devices through that network. IoT must ensure certain ML labels to confirm its trustworthiness for the device.

References

1. Aitken, M., Toreini, E., Carmichael, P., Coopamootoo, K., Elliott, K., & van Moorsel, A. (2020). Establishing a social licence for Financial Technology: Reflections on the role of the private sector in pursuing ethical data practices. *Big Data & Society, 7*(1), 2053951720908892.
2. Goodin, D., & Technica, A. (2015). 9 Baby monitors wide open to hacks that expose users' most private moments. [Online]. Available: http://tinyurl.com/ya7w43e9
3. Bhardwaj, A., Al-Turjman, F., Kumar, M., Stephan, T., & Mostarda, L. (2020). Capturing-the-Invisible (CTI): Behavior-based attacks recognition in IoT-oriented industrial control systems. *IEEE Access*, 1–1. https://doi.org/10.1109/ACCESS.2020.2998983
4. Hirsch, J., & Los Angeles Times. (2015). Hackers can now hitch a ride on car computers. [Online]. Available: http://www.latimes.com/business/autos/la-fi-hy-car-hacking-20150914-story.html
5. Atheron, K. D., & Popular Science. (2015). Hackers can tap into hospital drug pumps to serve lethal doses to patients. [Online]. Available: http://tinyurl.com/qfscthv
6. Pauli, D., & ITNews. (2015). Hacked terminals capable of causing pacemaker deaths. [Online]. Available: http://tinyurl.com/ycl4z9xf
7. Zhou, L., Wu, D., Zheng, B., & Guizani, M. (Mar. 2014). Joint physical-application layer security for wireless multimedia delivery. *IEEE Communications Magazine, 52*(3), 66–72.

8. Zhang, L., & Melodia, T. (2015, September). Hammer and anvil: The threat of a cross-layer jamming-aided data control attack in multihop wireless networks. In *Proceedings of the IEEE conference on communications and network security (CNS)*, Florence, Italy (pp 361–369).
9. Rathore, R., & Gau, C. (2014). Integrating biometric sensors into automotive Internet of Things. In *Proceedings of the IEEE international conference on cloud computing and Internet of Things (CCIOT)*, Changchun, China (pp. 178–181).
10. Laplante, P. A., & Laplante, N. (2016). The Internet of Things in healthcare: Potential applications and challenges. *IT Professional, 18*(3), 2–4.
11. Shankar, A., Pandiaraja, P., Sumathi, K., Stephan, T., & Sharma, P. (2020). Privacy preserving E-voting cloud system based on ID based encryption. *Peer-to-Peer Networking and Applications*. https://doi.org/10.1007/s12083-020-00977-4
12. Stephan, T., Al-Turjman, F., Joseph K, S., & Balusamy, B. (2020). Energy and spectrum aware unequal clustering with deep learning based primary user classification in cognitive radio sensor networks. *International Journal of Machine Learning and Cybernetics*. https://doi.org/10.1007/s13042-020-01154-y
13. Restuccia, F., Das, S. K., & Payton, J. (2016). Incentive mechanisms for participatory sensing: Survey and research challenges. *ACM Transaction on Sensor Networks, 12*(2), 1–40.
14. Mohammadi, V., Rahmani, A. M., Darwesh, A. M., et al. (2019). Trust-based recommendation systems in Internet of Things: A systematic literature review. *Human Centric Computing and Information Sciences, 9*, 21. https://doi.org/10.1186/s13673-019-0183-8
15. Shayesteh, B., Hakami, V., & Akbari, A. (2020). A trust management scheme for IoT-enabled environmental health/accessibility monitoring services. *International Journal of Information Security, 19*, 93–110. https://doi.org/10.1007/s10207-019-00446-x
16. Pudil, P., & Novovicová, J. (1998). Novel methods for feature subset selection with respect to problem knowledge. In *Feature extraction, construction and selection* (pp. 101–116). Springer.
17. Zhang, L., Restuccia, F., Melodia, T., & Pudlewski, S. M. (2017, October). Learning to detect and mitigate cross-layer attacks in wireless networks: Framework and applications. In *Proceedings of the IEEE conference on communications and network security (CNS)*, Las Vegas, NV, USA (pp. 361–369).
18. Cañedo, J., & Skjellum, A. (2016, December). Using machine learning to secure IoT systems. In *Proceedings of the 14th IEEE annual conference on privacy security trust (PST)*, Auckland, New Zealand (pp. 219–222).
19. Hofmeyr, S. A., & Forrest, S. (2000). Architecture for an artificial immune system. *Evolutionary Computation, 8*(4), 443–473.
20. Ning, H., & Liu, H. (2012). Cyber-physical-social based security architecture for future Internet of Things. *Advances in Internet of Things, 2*(1), 1–7.
21. Ning, H., Liu, H., & Yang, L. T. (2013). Cyberentity security in the Internet of Things. *Computer, 46*(4), 46–53.
22. Miettinen, M., et al. (2017, June). IoT Sentinel: Automated device-type identification for security enforcement in IoT. In *Proceedings of the IEEE 37th international conference on distributed computing systems (ICDCS)*, Atlanta, GA, USA (pp. 2177–2184).
23. Velliangiri, S., & Kasaraneni, K. K. (2020). Machine learning and deep learning in cyber security for IoT. In A. Kumar, M. Paprzycki, & V. Gunjan (Eds.), *ICDSMLA 2019* (Lecture notes in electrical engineering) (Vol. 601). Springer. https://doi.org/10.1007/978-981-15-1420-3_107
24. Chithaluru, P., Al-Turjman, F., Kumar, M., & Stephan, T. (2020). I-AREOR: An energy-balanced clustering protocol for implementing green IoT in smart cities. *Sustainable Cities and Society*, 102254. https://doi.org/10.1016/j.scs.2020.102254
25. Stephan, T., Al-Turjman, F., Joseph, K. S., Balusamy, B., & Srivastava, S. (2020). Artificial intelligence inspired energy and spectrum aware cluster based routing protocol for cognitive radio sensor networks. *Journal of Parallel and Distributed Computing*. https://doi.org/10.1016/j.jpdc.2020.04.007

26. Papernot, N., McDaniel, P., Sinha, A., & Wellman, M. P. (2018). Sok: Security and privacy in machine learning. In *2018 IEEE European symposium on security and privacy (EuroS&P)* (pp. 399–414). IEEE.

27. Eykholt, K., Evtimov, I., Fernandes, E., Li, B., Rahmati, A., Xiao, C., Prakash, A., Kohno, T., & Song, D. (2018, June). Robust physical-world attacks on deep learning visual classification. In *Conference on computer vision and pattern recognition*.

28. Al-Rubaie, M., & Chang, J. M. (2019). Privacy-preserving machine learning: Threats and solutions. *IEEE Security & Privacy, 17*(2), 49–58.

29. O'Neil, C. (2016). *Weapons of math destruction: How big data increases inequality and threatens democracy*. Crown Publishing Group.

30. Noble, S. U. (2018). *Algorithms of oppression: How search engines reinforce racism*. New York University Press.

31. Kearns, M. J., Neel, S., Roth, A., & Wu, Z. S. (2018). Preventing fairness gerrymandering: Auditing and learning for subgroup fairness. In *Proceedings of the 35th international conference on machine learning, ICML 2018* (pp. 2569–2577).

32. Liu, L., Dean, S., Rolf, E., Simchowitz, M., & Hardt, M. (2018). Delayed impact of fair machine learning. In *Proceedings international conference on machine learning* (Proceedings of machine learning research (PMLR)) (Vol. 80, pp. 3156–3164).

33. Bellamy, R. K. E., Dey, K., Hind, M., Hoffman, S. C., Houde, S., Kannan, K., Lohia, P., Martino, J., Mehta, S., Mojsilovic, A., Nagar, S., Ramamurthy, K. N., Richards, J. T., Saha, D., Sattigeri, P., Singh, M., Varshney, K. R., & Zhang, Y. (2018). AI fairness 360: An extensible toolkit for detecting, understanding, and mitigating unwanted algorithmic bias. *CoRR, abs/1810.01943*.

34. Doshi-Velez, F., Kortz, M., Budish, R., Bavitz, C., Gershman, S. J., O'Brien, D., Shieber, S., Waldo, J., Weinberger, D., & Wood, A. (2017). *Accountability of AI under the law: The role of explanation*. Berkman Center, Technical Report 18-07.

35. Koene, A., Clifton, C., Hatada, Y., Webb, H., & Richardson, R. (2019, April). *A governance framework for algorithmic accountability and transparency*. European Parliamentary Research Service, Technical Report PE 624.262.

36. Goodman, B., & Flaxman, S. R. (2017). European Union regulations on algorithmic decision-making and a "right to explanation". *AI Magazine, 38*(3), 50–57.

37. Wachter, S., Mittelstad, B., & Floridi, L. (Jun. 2017). Why a right to explanation of automated decision-making does not exist in the general data protection regulation. *International Data Privacy Law, 7*(2), 76–99.

38. Miller, T. (2019). Explanation in artificial intelligence: Insights from the social sciences. *Artificial Intelligence, 267*, 1–38.

39. Lipton, Z. C. (2018). The mythos of model interpretability. *Queue, 16*(3), 30:31–30:57.

40. Huysmans, J., Dejaeger, K., Mues, C., Vanthienen, J., & Baesens, B. (2011). An empirical evaluation of the comprehensibility of decision table, tree and rule based predictive models. *Decision Support Systems, 51*(1), 141–154.

41. Doshi-Velez, F. & Kim, B. (2017, February). Towards a rigorous science of interpretable machine learning. ArXiv e-prints.

42. Freedman, R., Schaich Borg, J., Sinnott-Armstrong, W., Dickerson, J. P., & Conitzer, V., (2018). Adapting a kidney exchange algorithm to align with human values. In *Proceedings of the 2018 AAAI/ACM conference on AI, ethics, and society, AIES '18* (pp. 115–115).

43. Graham, J., Meindl, P., Beall, E., Johnson, K. M., & Zhang, L. (2016). Cultural differences in moral judgment and behavior, across and within societies. *Current Opinion in Psychology, 8*, 125–130.

44. Awad, E., Dsouza, S., Kim, R., Schulz, J., Henrich, J., Shariff, A., Bonnefon, J.-F., & Rahwan, I. (2018). The moral machine experiment. *Nature, 563*(7729), 59.

45. Perreault, L. (2015). Big data and privacy: Control and awareness aspects. In *CONF-IRM, 15*.

46. Goodman, B., & Flaxman, S. (2016, June). European Union regulations on algorithmic decision-making and a "right to explanation". In *Proceedings of the 2016 ICML workshop on*

human interpretability in machine learning (WHI 2016). arXiv:1606.08813. http://arxiv.org/abs/1606.08813

47. Diakopoulos, N. (2016). Accountability in algorithmic decision making. *Communications of the ACM, 59*(2), 56–62. https://doi.org/10.1145/2844110

48. Burman, A. (2019). Will a GDPR-style data protection law work for India?.

49. Sacks, S. (2018, January 29). New China data privacy standard looks more far-reaching than GDPR. *Center for Strategic and International Studies.*

A Novel αβEvolving Agent Architecture for Designing and Development of Agent-Based Software

Shashank Sahu, Rashi Agarwal, and Rajesh Kumar Tyagi

1 Introduction

Software is developed to provide services to the customer. To develop software, the first requirements are gathered from the customer. These requirements are analysed and finalized by the analyst and it is forwarded to design and coding phase for implementation. As implementation is over, testing is performed on the software; it is then deployed at customer site for working. The customer starts using the software and gets outcome by executing the functionality of the software [1]. Whenever new requirements occur related to the functionality of the software, it is incorporated in the software using manual process [2]. Consider a case study: 'a retail software that provides a functionality of selling/purchasing of items. It consists of a list of items. Each item has its own price. Customer is purchasing an item on its price. Items' prices are not same all the time. Price changes with respect to time. These changes in the price act as a new requirement for the customer. These changes need to be incorporated in the retail software. To incorporate these changes, software needs to be updated, for example, in this case study, inclusion of 'latest price of an item in the retail software'. Currently, software incorporates new requirements

S. Sahu (✉)
Department of Computer Science & Engineering, Ajay Kumar Garg Engineering College, Ghaziabad, India
e-mail: sahushashank@akgec.ac.in

R. Agarwal
Department of Master of Computer Applications, Galgotias College of Engineering & Technology, Greater Noida, India
e-mail: rashi.agarwal@galgotiacollege.edu

R. K. Tyagi
Department of Computer Science & Engineering, Amity University, Gurugram, India
e-mail: rktyagi@ggn.amity.edu

© The Author(s), under exclusive license to Springer Nature Switzerland AG 2022
F. Al-Turjman et al. (eds.), *Transforming Management with AI, Big-Data, and IoT*, https://doi.org/10.1007/978-3-030-86749-2_10

of customer manually [3]. In the manual process, software provides facility to customer for inclusion of new changes/requirements that occur in the functionality, and customer includes these changes as and when required through facility provided in the software. Consider the case study of retail software in which price of item changes frequently and it is also fluctuates with respect to time [4]. Customer includes these changes of price in the retail software through facility given in the software [5]. Inclusion of new requirements/changes in the functionality becomes a tedious job for the customer, and it also needs extra time to reflect latest update in the functionality [6]. We need a solution to overcome this problem. The solution should be able to accommodate changes that occur over a period of time without human intervention. The key advantage intelligent agents add to the technologies like IoT [42, 43] and other artificial intelligence systems [44] is developing a system based on users' needs. Software evolving agent is capable of solving this problem. Software evolving agent observes changes that occur as time moves ahead and incorporates these changes in the software itself without human intervention. The chapter proposes an architecture of software evolving agent which is capable of incorporating changes without human intervention. The proposed evolving agent evolves itself with respect to time, and there is no need for any human intervention. The chapter's layout is as follows. Section "Literature Review" discusses background related to retail shop and evolving software agent. Section "αβEvolving Agent (αβEA) Architecture" presents proposed work. Section "Experiment" describes experiments performed on the proposed work. Section "Result" discusses the results obtained from the experiments carried out. Section "Conclusion" presents the conclusion of the work.

2 Literature Review

The goal of any retail software is to provide facilities of selling and purchasing of items [7]. The list of items is maintained in the database of the software. Price of each item is also stored in the database for selling and purchasing [8]. Retail software [45] helps customer for easy billing and fast processing of items to be purchased by customer. Today, all supermarkets are using retail software [9]. Software agent is a software component that observers change in the requirements happening in the environment and reacts accordingly without any human intervention [10].

A software agent performs the tasks of other entities like software, hardware or a human. An agent reflects intelligence and has autonomy. An agent senses the environment, and on the basis of environment condition, it gives response to the environment [10]. Therefore, the agent is able to take actions itself according to the changes in the environment. This agent's ability makes it differ from simple software entity which performs only predefined tasks. An agent has autonomy, reactivity, pro-activeness and social abilities. Agents can learn the requirements of users themselves and accordingly fulfil requirements of the user. The software system which is developed using agent technology may have many agents. All agents perform various tasks of the system in coordination and collaboration. A software agent consists of properties like autonomy, reactivity, pro-activeness, social ability.

A software agent consists of properties like autonomy, reactivity, pro-activeness, social ability. Autonomy provides the ability of taking decision. Reactivity is the ability of reacting according to changes that occur in the environment. Pro-activeness gives the ability of thinking in advance for problems that may arise in near future and accordingly prepares itself to handle those problems [11]. A new paradigm of programming is emerging which is called Agent-Oriented Software Engineering (AOSE) [12]. Software component in this paradigm of programming is considered as software agent which consists of agent's properties as discussed. AOSE software development consists of set of software agents which work actively to achieve the desired goal. To design software based on software agents, it is necessary to have an architecture of software agents [13]. Computer programs that are able to evolve themselves have been advocated in [14–16]. The evolution of programs may be realized using software agents, and agents are able to take suitable decision for its evolution [17]. Evolution of computer program needs development of adaptable architecture, which is also called evolving architecture. Evolving architecture is able to handle changes happening in the dynamic environment [18]. Software development using software agents is called agent-oriented software development. Programming for software agents need a suitable architecture on which software agents can be developed. Some architecture has been proposed in [19–22]. A road map related to various architecture of software agents have been discussed in [16]. However, a suitable architecture is needed to design software agents that evolve itself [16]. Changes in the retail price of items reflect in retail margins [23]. Price of items in supermarket is changing on weekly basis. Generally, price of items is more at the start of the week and it drops down up to 50% during end of the week [23]. Price of items also varies with respect to time of the day. Price of items increases during peak time of day and price reduces in normal timings. Change in the prices also increases the effort of changing price labels on the product [24]. Quantity demand of certain items is also affected by change in the price [25–26]. Grocery industry has repetitive and manual labour work. It needs automation to reduce this labour work [27].

Software agents are used in marketing of supermarkets in [28]. A personalized recommender system has been designed which suggests new products to customers of supermarkets [29]. Architecture of evolving agent is proposed in [30], and it is experimented on an application of purchasing items. An agent is proposed that provides a list of prices of items to customers for purchasing. It also provides special discounts available on the item. The proposed agent facilitates user hassle-free purchasing of items [31]. A self-evolving scheme is presented in [32] based on agents without human intervention for simulation. Use of agents as a middle ware for server side and client side for performance is presented in [33]. Software agents are used for identification, assessment and monitoring of risks agile software [34]. An interactive adaptive multiagent tool has been developed that evolves ontology from text [35]. A framework based on agent is proposed to simulate different health states and behaviour of persons to predict epidemic automatically [36]. Authors in [37–41] discussed how multiagent systems help in self-adaptive systems.

The agent's design and its development is different from traditional software development. Therefore, a separate paradigm is needed for evolving agent development. Evolving agent is used for solving modern complex problem. Applications of agent software developments are naval warship, customer satisfaction, e-commerce, retail shops and distributed computing. There are various frameworks of agent-oriented requirements engineering, and these frameworks are evaluated using various parameters. Authors emphasized that requirements of customers are dynamic. These types of requirements that are changing frequently can be addressed using agents. A spiral model framework for requirements engineering is proposed that uses user story cards to specify the requirement [42]. Developing the software using agents is a new era of software development. Goals and tasks can be assigned to agent. It is the agent's responsibility to accomplish its assigned tasks. The agent is also responsible to evolve itself because of new changes occurring in the environment and take suitable decisions for better results in near future [43]. Architecture for web services is presented in [44]. It is written in CAMLE modelling language and specification using SLABS. The usage of agents in web services applications was presented in this paper. The requirement of goal and protocol for the evolving agent is addressed using interaction diagram in [45]. In this paper, the authors concluded that addition of goal in the evolving agent is both usable and effective. Agents may perform various roles for accomplishing the designated task. The agent can be developed using component-based architecture [46]. Pour [47] emphasized that evolving software development using agents is an efficient and effective concept. Agents can be used in vehicle management processes like tracking, detection and control of vehicles at the intersections [48–49]. Intelligent agents are changing the development approach of various applications specifically in solving the modern complex problems [50, 51].

3   αβEvolving Agent (αβEA) Architecture

This section describes proposed architecture αβEvolving Agent (αβEA) which is presented in Fig. 1. This architecture provides basis for designing the evolving agent. The architecture consists of five components which are listed below:

(i) Learning
(ii) α factor
(iii) β factor
(iv) Matured memory
(v) Pre-matured memory

The property of evolving agent is to adapt the changes of environment itself. It does not involve intervention of human being during its evolution. This architecture fulfils this property and defines the structure of the evolving agent which makes the

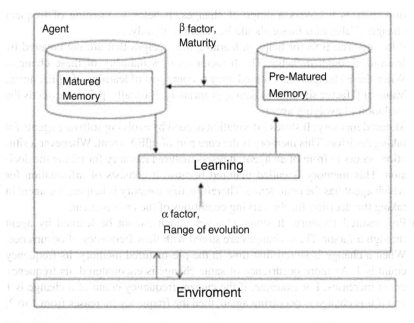

Fig. 1 Evolving Agent αβEA (αβEvolving Agent)

agent adaptable. The evolving agent developed using this architecture can be implemented easily using any programming language. The major bottleneck regarding evolving agent is its implementation. Developers are facing problem of how to implement an agent which is able to be evolved. The proposed architecture solves this implementation problem. This architecture has five components which work together to give capability of adaptation to evolving agent.

Each component of proposed architecture works individually and provides information to respective component. Working of the evolving agent is controlled by both α and β factors. The memory components provide the memorising capability to the agent. The architecture has a learning component which enables evolving agent to take suitable decision based on the changes in the environment. Data is moving from one component to another component. Evolving agent based on the proposed architecture senses the external environment, and the information gathered is processed according to α and β factors. The first information is moved to pre-matured memory, then it is moved to matured memory on basis β factor. Following is the description of each component.

(i) Learning: Learning component observes the changes that are occurring in the environment related to functionality of the software. It learns from these changes, and evolving agent αβEA evolves itself through this learning. Agent αβEA uses two factors for learning: α factor and β factor.

(ii) α factor: The α factor covers range of changes that are likely to occur in the functionality of the software in the environment. The α factor is called evolution span as it covers a range of changes. It helps in learning of frequent changes. Value of α factor should be selected suitably.

(iii) β factor:- The β factor helps in learning the changes that are not covered by learning through α factor. The β factor ensures maturity of these changes. When these changes are matured, they become part of learning of αβEA agent. Value of β factor should be taken appropriately. Basically, β factor extends the evolution of evolving agent.

(iv) Matured memory: It stores information needed by evolving software agents for taking decision. This memory is the core part of αβEA agent. Whenever a situation occurs in front of an agent, it refers matured memory for taking the decision. This memory is called matured because it consists of information for which agent has the confidence. Therefore, this memory is helping the agent in taking the decision for the varying conditions of the environment.

(v) Pre-matured memory: It stores changes which cannot be learned by agent through α factor. These changes are stored with their frequency of occurrence. When a change is stored first time in the pre-matured memory, its frequency count is 1. As more occurrence of same change is encountered, its frequency count increases. For example, if the current frequency count of a change is 1 and if this change is occurring again, then the frequency increases from 1 to 2.

3.1   Working of αβEA Agent

This section describes the working of proposed agent αβEA architecture. Flowchart of working is given in Fig. 2. Proposed agent αβEA receives input from the user. It then uses matured memory to understand the input received. When the agent finds that the input received from the outside world matches with information present in the matured memory, it takes the decision based on the input. If input does not match with the information present in the matured memory, then it shows that input received from user is different from the information present in the matured memory. It also shows that some change has happened in the input as compared to previous input. We call this input as *change input*. Evolving agent αβEA is required to evolve itself to adapt these changes and able to take better decision for the new situation. To understand what types of changes are in the change input, evolving agent αβEA follows two strategies. These strategies work sequentially. These strategies are described below:

(a) First, the agent uses α evolution span factor to find out that whether changes in the change input is within the range of α evolution span. If it is within the range of evolution span, evolving agent understands that there is small change in the change input in comparison with the information present in matured memory. In this case, evolving agent does not store any change of this change input in the matured memory. However, evolving agent is able to take decision on this change input according to policy which has been established.

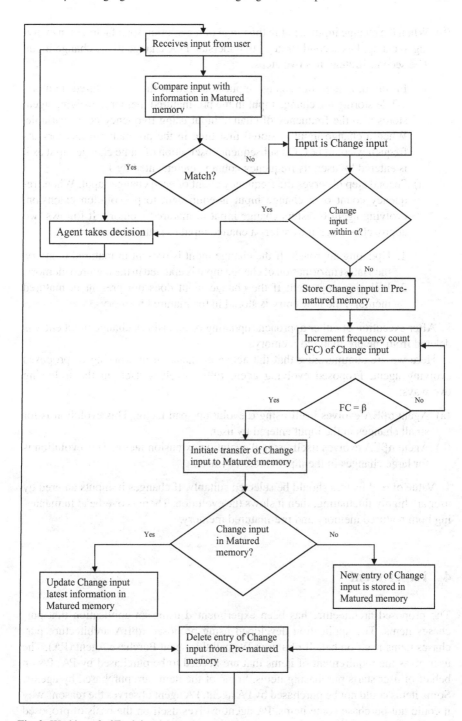

Fig. 2 Working of αβEvolving Agent

(b) When the change input is not in the range of α evolution span factor, then evolving agent applies second strategy to understand the change in the change input. The second strategy has two steps:

(i) In this first step, the change input is stored in the pre-matured memory. While storing the change input in the pre-matured memory, evolving agent stores also the frequency of change input using frequency count variable. When a change input is stored first time in the pre-matured memory, its frequency count is 1. On subsequent observation of same change input as it is entered by user, its frequency count is incremented by 1.

(ii) Second step observes the frequency count of each change input. When frequency count of a change input is equivalent to β evolution extension, evolving agent transfers change input to matured memory. It follows two approaches when it transfers a change input to matured memory.

1. Updating approach: If the change input is present in matured memory, then latest information of change input is updated in the matured memory.
2. Inclusion approach: If the change input does not present in matured memory, the new entry is stored in the matured memory.

After execution of either approach (updating or inclusion), change input entry is deleted from the pre-matured memory.

Here we are summarizing that the above explanation of working of proposed evolving agent. Proposed evolving agent αβEA evolves itself in the following two ways:

(a) Agent αβEA evolves itself using α evolution span factor. This evolution is for small changes in the input entered by user.
(b) Agent αβEA evolves itself using β evolution extension factor. This evolution is for large changes in the input entered by user.

Value of α, β factors should be selected suitably. If changes in inputs entered by user are highly fluctuating, then it slows the evolution. There is overhead in managing both matured memory and pre-matured memory.

4 Experiment

The proposed architecture has been experimented using an application that purchases items. The application developed using proposed αβEA architecture purchases items itself on behalf of a user. We call this agent Purchase Agent (PA). The user gives the requirement of items that are needed to be purchased by PA. PA on behalf of user starts purchasing items. Some of the items are purchased by agents. Some items could not be purchased by PA agent. PA agent observes the reasons why it could not purchase some items. PA agent evolves itself on the basis of proposed αβEA strategies so that it can purchase more items in near future. This way PA evolves itself by adapting new changes happening in its functionality timely. Evolution of PA agent enables PA to take better decision in the near future. Evolution of PA does not involve any human intervention. PA agent is written in JSP (Java Sever pages) language which consists of more than 500 lines. Matured and Prematured repositories are maintained in MS Access database. Evolution factors α and β are used as variables in PA agent program. Experiments are performed on more than 12,000 datasets on various combinations of β and α values where β = 1, 2, 3 and α = 10%, 20%, 30%, 40%, 50%. Figure 3 shows 15 experiments on different combinations of β and α. Experiments have been performed on Intel dual core 2.1 GHz processor, 2 GB RAM computer system. Experiments are performed on the following factors:

A. Number of items purchased by evolving system (PA agent)
B. Execution time of evolved system and non-evolved system
C. Storage requirement for evolved system and non-evolved system
D. Number of interactions needed in evolved system and non-evolved system

A. Figure 3 shows result of experiments. It shows the total number of items purchased by PA agent. Figure 3 has six columns. The first column shows the value of β and α, second column shows number of modifications required for evolution of PA agent. The third column shows items purchased because of α span factor. The fourth column shows value of third column in the percentage. The fifth column shows total items purchased by PA agent using both factors β and α. Sixth column shows values of fifth column in the percentage. Starting of Fig. 3 shows size of dataset used in the experiments and number of modifications required in non-evolved system.

Data Set size	12932
Non-evolved Total modifications	395

	Evolved Total Modification	Items Purchased with α span	%	Total Items Purchased	%
β=1 α=10%	136	6100	47.17	12795	98.94
β=1 α=20%	68	9022	69.76	12863	99.47
β=1 α=30%	48	9945	76.90	12883	99.62
β=1 α=40%	39	10087	78.00	12892	99.69
β=1 α=50%	28	11178	86.44	12903	99.78

	Evolved Total Modification	Items Purchased with α span	%	Total Items Purchased	%
β=2 α=10%	130	5998	46.38	12669	97.97
β=2 α=20%	65	9009	69.66	12798	98.96
β=2 α=30%	43	9945	76.90	12841	99.30
β=2 α=40%	38	10068	77.85	12852	99.38
β=2 α=50%	27	11173	86.40	12875	99.56

	Evolved Total Modification	Items Purchased with α span	%	Total Items Purchased	%
β=3 α=10%	123	5877	45.45	12550	97.05
β=3 α=20%	66	9029	69.82	12737	98.49
β=3 α=30%	42	9817	75.91	12799	98.97
β=3 α=40%	38	10067	77.85	12813	99.08
β=3 α=50%	27	11173	86.40	12848	99.35

Fig. 3 Items purchased by PA agent

Experiment shows success of purchasing of items from 97.05% to 99.78% by PA agent with self-evolution.

B. Execution time of evolved system is compared with execution time of non-evolved system. Figure 4 shows this comparison. Comparison shows saving of 84.08% in execution time.

C. Evolved and non-evolved system is compared with respect to storage requirement of evolved and non-evolved system. A record occupies 504 bytes of space. For non-evolved system, a total of 12,932 records are needed and 6,517,728 bytes of storage are needed. In case of evolved system, the number of records is reduced to 9232 records and an average saving of 28.61% storage. For storage in terms of bytes, it is reduced to 4,616,507 bytes and average saving of 29.17% (Fig. 5).

D. Evolved and non-evolved system is compared with respect to number of interactions needed for updating in the storage. In non-evolved system, a total of 395 interactions are needed. In case of evolved system, on average, 61.225 interactions are needed. It shows saving in interactions on an average of 84.5% (Fig. 6).

5 Result

A comparison of various parameters of the experiments is shown from Figs. 7, 8, 9, 10 and 11. Result summary of the experiments is shown in Fig. 12. Figure 7 shows an increment of 2.84% in purchasing of items using evolving PA agent which is based on proposed $\alpha\beta$EA architecture. Figure 8 shows execution time of purchasing items is drastically reduced to 84.08% because of using evolving PA agent. Figures 9 and 10 show a reduction in storage requirement. A saving of 28.61% is observed in storage requirement in terms of records, and a saving of 29.17% is observed in

Non-Evolved System	
	1577 Sec
	26 Min

	Evolved System							
	$\alpha = 10\%$	$\alpha = 20\%$	$\alpha = 30\%$	$\alpha = 40\%$	$\alpha = 50\%$	Avg	Avg Saving	Avg Saving %
$\beta=1$	96	107	106	104	109	104.4	1472.6	93.38
	1.6	1.7	1.7	1.7	1.8	1.7	24.3	
$\beta=2$	433	385	299	143	118	275.6	1301.4	82.52
	7.2	6.4	4.9	2.3	1.9	4.54	21.46	
$\beta=3$	498	435	345	468	120	373.2	1203.8	76.33
	8.3	7.2	5.7	7.8	2	6.2	19.8	

* First cell represents value in seconds
Second cell represents value in Minutes

Fig. 4 Comparing execution time of evolved and non-evolved systems

Non-Evolved System	Records	Bytes
(504 bytes /Record)	**12932**	**6517728**

	Evolved System								
	α= 10%	α= 20%	α= 30%	α= 40%	α= 50%	Avg	Avg Saving	Avg Saving %	Avg Saving %
β= 1	6100	9022	9945	10087	11178	9266.4	3665.6	28.35	28.63
	3050000	4511000	4972500	5043500	5589000	4633200	1884528	28.91	
β= 2	5998	9009	9945	10068	11173	9238.6	3693.4	28.56	28.84
	2999000	4504500	4972500	5034000	5586500	4619300	1898428	29.13	
β= 3	5877	9029	9817	10067	11173	9192.6	3739.4	28.92	29.20
	2938500	4514500	4908500	5033500	5586500	4596300	1921428	29.48	

* First cell represents value in no. of Records
  Second cell represents value in Bytes

Fig. 5 Comparing storage requirement of evolved and non-evolved systems

Non-Evolved System	No of interactions
	395

	Evolved System					
	α = 10%	α = 20%	α = 30%	α = 40%	α = 50%	Avg Saving %
β=1	136	68	48	39	28	83.85
	259	327	347	356	367	
	65.57	82.78	87.85	90.13	92.91	
β= 2	130	65	43	38	27	84.66
	265	330	352	357	368	
	67.09	83.54	89.11	90.38	93.16	
β= 3	123	66	42	38	27	85.01
	272	329	353	357	368	
	68.86	83.29	89.37	90.38	93.16	

* First cell represents no. of interactions
  Second cell represents Saving (in no. of interactions)
  Third cell represents Saving (in %)

Fig. 6 Comparing number of interactions for updating in evolved and non-evolved systems

storage requirement in terms of bytes. A saving of 84.5% is observed which is significantly high in number of interactions required for updating the records in storage in Fig. 11. Figure 12 shows result summary of the experiments. It shows that the use of proposed evolving αβEA architecture has higher impact on execution time and number of interactions; at the same time, it also has a significant impact on storage

Fig. 7 Comparison of items purchased

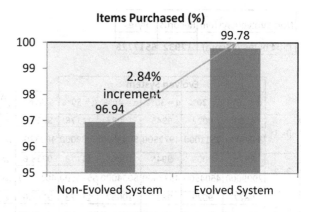

Fig. 8 Comparison of execution time

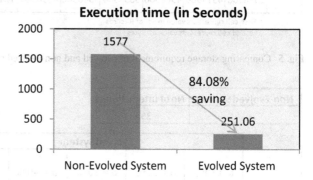

requirement. Result summary also shows that a number of items can be purchased by software agent without involvement of human beings with increased purchasing. A software agent evolves itself by accommodating changes occurring in the functionality time to time without intervention of human beings for the better decision for new situation.

6 Conclusion

The proposed evolving αβEA architecture is suitable for designing a software agent. Architecture also supports evolving capability in the agent. Evolution in the software agents is highly needed today because of functionality of software changes as the time passes. Generally, these changes are incorporated in the software with help of human beings. But the proposed agent αβEA is able to incorporate these changes itself and evolves itself without the intervention of human beings. As an experiment, a software agent named Purchase Agent (PA) has been designed using proposed agent αβEA architecture and implemented in JSP language to find out its effectiveness. Experiments show that it is possible to design and develop such software agent

Fig. 9 Comparison of storage (in records)

Storage (in Records)

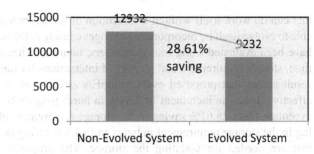

28.61% saving

Non-Evolved System   Evolved System

Fig. 10 Comparison of storage (in bytes)

Storage (in Bytes)

6517728

29.17% saving

4616507

Non-Evolved System   Evolved System

Fig. 11 Comparison of number of interactions for updating

No. of interactions

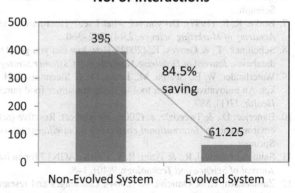

395

84.5% saving

61.225

Non-Evolved System   Evolved System

Result (in %)	
Number of items purchased	2.84% increment
Execution time	84.08% saving
Storage requirement (in Records)	28.61% saving
Storage requirement (in Bytes)	29.17% saving
Number of interactions for updation	84.5% saving

Fig. 12 Result summary of experiments

that can do work itself without intervention of human beings and the agent is also able to evolve itself by incorporating changes coming in the near future. Experiments have been evaluated using four parameters: number of items purchased, execution time, storage requirement and number of interactions for updating. The experiment result shows that proposed evolving $\alpha\beta$EA architecture is highly promising and effective. It has an increment of 2.84% in purchasing items, 84.05% saving in the execution time, 28.61% saving in the storage requirement of records, 29.17% saving in the storage requirement in bytes and 84.5% saving in number of interactions that are needed for updating the storage. The proposed agent may be further extended to include more factors that may help the agent evolution.

References

1. Bosch, J. (2000). *Design and use of software architectures: Adopting and evolving a product-line approach*. Pearson Education.
2. German, D. M. (2006). An empirical study of fine-grained software modifications. *Empirical Software Engineering, 11*(3), 369–393.
3. Gefen, D., & Schneberger, S. L. (1996, November). The non-homogeneous maintenance periods: A case study of software modifications. In *ICSM* (pp. 134–141). IEEE.
4. Gardner, B. L. (1975). The farm-retail price spread in a competitive food industry. *American Journal of Agricultural Economics, 57*(3), 399–409.
5. Levy, M., Grewal, D., Kopalle, P. K., & Hess, J. D. (2004). Emerging trends in retail pricing practice: Implications for research. *Journal of Retailing, 80*(3), xiii–xx.
6. Grubb, P., & Takang, A. A. (2003). *Software maintenance: Concepts and practice*. World Scientific.
7. Burke, R. R. (1997). Do you see what I see? The future of virtual shopping. *Journal of the Academy of Marketing Science, 25*(4), 352–360.
8. Schellinck, T., & Groves, K. (2002). How low can you go? The value of sparse data in retail databases. *Journal of Database Marketing & Customer Strategy Management, 9*(2), 143–149.
9. Waterlander, W. E., Scarpa, M., Lentz, D., & Steenhuis, I. H. (2011). The virtual supermarket: An innovative research tool to study consumer food purchasing behaviour. *BMC Public Health, 11*(1), 589.
10. Banerjee, D., & Tweedale, J. (2006, September). Reactive (re) planning agents in a dynamic environment. In *International conference on intelligent information processing* (pp. 33–42). Springer.
11. Sahu, S., Agarwal, R., & Tyagi, R. K. (2016). AGNT7 for an intelligent software agent. *Indian Journal of Science and Technology, 9*(40), 1–8.
12. Zambonelli, F., & Omicini, A. (2004). Challenges and research directions in agent-oriented software engineering. *Autonomous Agents and Multi-Agent Systems, 9*(3), 253–283.
13. Jennings, N. R. (2000). On agent-based software engineering. *Artificial Intelligence, 117*(2), 277–296.
14. Rosenblatt, F. (1958). The perceptron: A probabilistic model for information storage and organization in the brain. *Psychological Review, 65*(6), 386.
15. Samuel, A. L. (1959). Some studies in machine learning using the game of checkers. *IBM Journal of Research and Development, 3*(3), 210–229.
16. Fogel, D. B. (2006). *Evolutionary computation: Toward a new philosophy of machine intelligence* (Vol. 1). Wiley.
17. Hanna, L., & Cagan, J. (2008, January). Evolutionary multi-agent systems: An adaptive approach to optimization in dynamic environments. In *ASME 2008 international design*

engineering technical conferences and computers and information in engineering conference (pp. 479–487). American Society of Mechanical Engineers.

18. Nunes, I., Nunes, C., Kulesza, U., & Lucena, C. (2008, May). Developing and evolving a multi-agent system product line: An exploratory study. In *International workshop on agent-oriented software engineering* (pp. 228–242). Springer.

19. Bratman, M. E., Israel, D. J., & Pollack, M. E. (1988). Plans and resource-bounded practical reasoning. *Computational Intelligence, 4*(3), 349–355.

20. Doyle, J. (1992). Rationality and its roles in reasoning. *Computational Intelligence, 8*(2), 376–409.

21. Rao, A. S., & Georgeff, M. P. (1991). Modeling rational agents within a BDI-architecture. *KR, 91*, 473–484.

22. Shoham, Y. (1993). Agent-oriented programming. *Artificial Intelligence, 60*(1), 51–92.

23. Hosken, D., Matsa, D., & Reiffen, D. A. (2000). *How do retailers adjust prices?: Evidence from store-level data.* Federal Trade Commission.

24. Ball, K., McNaughton, S. A., Mhurchu, C. N., Andrianopoulos, N., Inglis, V., McNeilly, B., ... Crawford, D. (2011). Supermarket Healthy Eating for Life (SHELf): Protocol of a randomised controlled trial promoting healthy food and beverage consumption through price reduction and skill-building strategies. *BMC Public Health, 11*(1), 715.

25. Waterlander, W. E., Steenhuis, I. H., de Boer, M. R., Schuit, A. J., & Seidell, J. C. (2012). The effects of a 25% discount on fruits and vegetables: Results of a randomized trial in a three-dimensional web-based supermarket. *International Journal of Behavioral Nutrition and Physical Activity, 9*(1), 11.

26. González, X., & Miles-Touya, D. (2018). Price dispersion, chain heterogeneity, and search in online grocery markets. *SERIEs, 9*, 115–139.

27. Web site https://pathoverblog.wordpress.com

28. Rykowski, J. (2005). Active advertisement in supermarkets using personal agents. In *Challenges of expanding internet: E-commerce, E-business, and E-government* (pp. 405–419). Springer.

29. Lawrence, R. D., Almasi, G. S., Kotlyar, V., Viveros, M., & Duri, S. S. (2001). Personalization of supermarket product recommendations. In *Applications of data mining to electronic commerce* (pp. 11–32). Springer.

30. Sahu, S., Agarwal, R., & Tyagi, R. K. (2017). A novel OLDA evolving agent architecture. *Journal of Statistics and Management Systems, 20*(4), 553–564.

31. Benedicenti, L., Chen, X., Cao, X., & Paranjape, R. (2004, May). An agent-based shopping system. In *Electrical and computer engineering, 2004. Canadian conference on* (Vol. 2, pp. 703–705). IEEE.

32. Kang, D. O., Bae, J. W., & Paik, E. (2016, December). Incremental self-evolving framework for agent-based simulation. In *Computational science and computational intelligence (CSCI), 2016 international conference on* (pp. 1428–1429). IEEE.

33. Ivanović, M., Vidaković, M., Budimac, Z., & Mitrović, D. (2017). A scalable distributed architecture for client and server-side software agents. *Vietnam Journal of Computer Science, 4*(2), 127–137.

34. Odzaly, E. E., Greer, D., & Stewart, D. (2017). Agile risk management using software agents. *Journal of Ambient Intelligence and Humanized Computing, 9*, 823–841.

35. Sellami, Z., Camps, V., & Aussenac-Gilles, N. (2013). DYNAMO-MAS: A multi-agent system for ontology evolution from text. *Journal on Data Semantics, 2*(2–3), 145–161.

36. Miksch, F., Urach, C., Einzinger, P., & Zauner, G. (2014, April). A flexible agent-based framework for infectious disease modeling. In *Information and communication technology-EurAsia conference* (pp. 36–45). Springer.

37. Weyns, D., & Georgeff, M. (2010). Self-adaptation using multiagent systems. *IEEE Software, 27*(1), 86–91.

38. Iliadis, L., & Badica, C. (2020). Evolving and intelligent systems applications special issue. *Evolving Systems, 11*, 199–200.

39. Zhou, H., & Hirasawa, K. (2019). Evolving temporal association rules in recommender system. *Neural Computing & Applications, 31,* 2605–2619.
40. Sulis, E., Terna, P., Di Leva, A., et al. (2020). Agent-oriented decision support system for business processes management with genetic algorithm optimization: An application in healthcare. *Journal of Medical Systems, 44,* 157.
41. Djurdjevac Conrad, N., Helfmann, L., Zonker, J., et al. (2018). Human mobility and innovation spreading in ancient times: A stochastic agent-based simulation approach. *EPJ Data Science, 7,* 24.
42. Gaur, V., Soni, A., & Bedi, P. (2010). An agent-oriented approach to requirements engineering. In *2010 IEEE 2nd international advance computing conference (IACC)* (pp. 449–454). IEEE.
43. Gaur, V., Soni, A., & Bedi, P. (2010). An application of multi-person decision-making model for negotiating and prioritizing requirements in agent-oriented paradigm. In *DSDE 2010 – International conference on data storage and data engineering* (pp. 164–168). IEEE.
44. Zhu, H., & Shan, L. (2005). Agent-oriented modelling and specification of Web services. In *Proceedings – International workshop on object-oriented real-time dependable systems, WORDS* (pp. 152–159). IEEE.
45. Becker, J., Uhr, W., & Vering, O. (2001). Distributed retail information systems (DRS). In *Retail information systems based on SAP products. SAP excellence.* Springer.
46. Khallouf, J., & Winikoff, M. (2005). Towards goal-oriented design of agent systems. In *Proceedings – International conference on quality software* (Vol. 2005, pp. 389–394). IEEE.
47. Qu, Y., Wang, C., Lili Zhong, H. Z., & H. L. (2009). Research for an intelligent component-oriented software development approaches. *Journal of Software, 4*(10), 1136–1144.
48. Pour, G. (2002). Integrating agent-oriented enterprise software engineering into software engineering curriculum. *32nd Annual Frontiers in Education, 3,* 8–12.
49. Yadav, S. P. (2020). Vision-based detection, tracking and classification of vehicles. *IEIE Transactions on Smart Processing and Computing, 9*(6), 427–434, SCOPUS, ISSN: 2287–5255.
50. Sahu, S., Agarwal, R., & Tyagi, R. K. (2019). Fuzzy vehicle control system for single intersection. *International Journal of Recent Technology and Engineering (IJRTE), 8*(2S7). ISSN: 2277–3878.
51. Yadav, S. P., Mahato, D. P., & Linh, N. T. D. (2020). *Distributed artificial intelligence: A modern approach* (1st ed.). CRC Press.

Software-Defined Network (SDN) for Cloud-Based Internet of Things

Charu Awasthi, Isha Sehgal, Pawan Kumar Pal, and Prashant Kumar Mishra

1 Introduction

With the emergence of IoT (Internet of Things) [1], the number of devices connected to the Internet has grown exponentially and is expected to grow more in the near future. Because of users' changing demand due to variety of online services, the devices connected to the Internet are heterogeneous. These heterogeneous devices are connected to network and distributed globally all over the world covering the remote terrains as well. Every minute new users are connected to the Internet, existing users either keep on changing the devices or might switch to a different network. Usage of IoT devices is shown in Fig. 1. With the advent of new applications every single time, network traffic is getting increased day by day and thereby increasing the network load and changing the characteristics of the existing network. Traffic characteristics like traffic congestion, bandwidth, delay etc. are continuously changing, thereby demanding the reconfiguration of the network every single time to handle such changes. Therefore, the network needs to be modified accordingly to serve such applications.

Existing traditional technologies cannot handle such vast and ever-changing network requirements. Therefore, we require an intelligent network infrastructure like Software-Defined Network (SDN) [1, 2] which can cater to the needs of distributed and vast network and handle the heterogeneity of the network devices and applications without deteriorating the network performance [1].

C. Awasthi (✉) · P. K. Pal · P. K. Mishra
Department of Computer Science and Engineering, PSIT, Kanpur, Uttar Pradesh, India
e-mail: charu.awasthi@psit.ac.in

I. Sehgal
Department of Information Technology, PSIT College of Engineering,
Kanpur, Uttar Pradesh, India

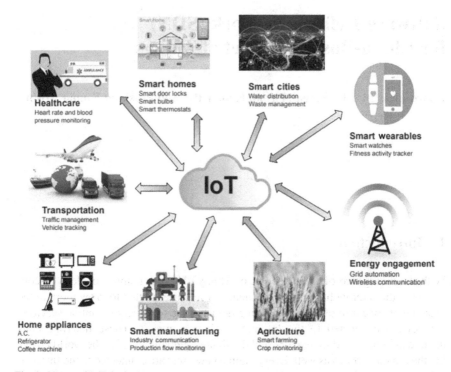

Fig. 1 Usage of IoT devices

It is already known that users get connected to the network and use its services if they have subscribed for network service which is made available by a service provider. The present scenario says that there are tons of users connected to multiple network service providers. It is predicted that the number of users connected to network will increase way more, thereby raising the load of service providers. This may suggest that each service provider should deploy a separate network for supporting its subscribers, thereby procuring resources and making huge investment for the same. If each one of the service providers deploys separate networks to cater the changing needs of the users and the network, then it would be too costly for the service providers and is neither practically possible. So, the network infrastructure should be reliable yet cost-effective.

The above problems suggest the need of the network infrastructure which is flexible to the ever-changing needs and can support the sharing of resources. This suggests that there can be a single network infrastructure and multiple service providers can share it. With sharable network infrastructure, the service provider can lease in the resources depending on the requirements of the users. With this technique, same network infrastructure resources can be shared and used by multiple service providers without requiring them to procure separate network and network elements. However, to support critical applications and requirements, the shared infrastructure must allow service providers to reserve the resources for crucial situations.

With the changing traffic and usage of shared network, it is required to change the existing security policies and provisioning services to secure the network [3] used by subscriber of any service provider. Specifically, policies related to network traffic generated by such newly introduced applications needs to be reconfigured. This means that high-level processing involving packet handling policies must be reconfigured to support this newly added traffic. Reconfiguring the network involves changing the location and policies of network elements as well like gateways, firewalls, load balancers and so on. With shared infrastructure, reconfiguring the network every time will be very difficult and not feasible keeping in mind that multiple service providers have shared the same resource. Therefore, there must be the provision of isolation of part of the network and resources when it is allotted to any service provider. This isolation requirement can be accomplished with the newly emerged technologies, that is, Software-Defined Networks (SDNs) [3, 4] and Network Function Virtualization (NFV). This chapter investigates these futuristic technologies, that is, SDN and NFV [3, 5] and will also provide the comparative study of these technologies with the traditional technologies. This integration of SDN and NFV leads to few open innovations like isolation of network [3] by utilizing the properties of SDN and NFV as shown in Fig. 2.

The history of the SDN principles, how this technology evolved and is getting collaborated with another technology, that is, NFV, also needs to be investigated for integration with IoT cloud [6]. The reason behind the adoption of SDN is its property of making network flexible and dynamic will work brilliantly on the problems being faced by traditional ways and are potentially solved by SDN to a greater extent. The SDN architecture comprises of three layers, that is, application layer, control plane and data plane along with the interfaces and SDN controller. SDN controller is an centralized entity that manages the working of SDN and NFV. NFV

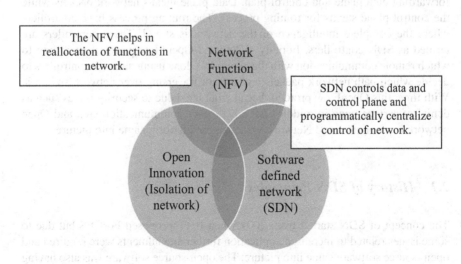

The NFV helps in reallocation of functions in network.

Network Function (NFV)

SDN controls data and control plane and programmatically centralize control of network.

Open Innovation (Isolation of network)

Software defined network (SDN)

Fig. 2 SDN and NFV contribution to open innovation

architecture is composed of two planes for management and controlling purpose; it is composed of VNF software which comprises of hardware layer, hypervisor layer and VM layer. Relationship of SDN and NFV can be explained in a single line "NFV does not need SDN it complements SDN" [6] which means both combine to produce better and reliable network against traffic steering. Cloud computing-based IoT is considered as IoT devices can also collect large amounts of data (big data) and store it on related cloud for further usage. Cloud computing and its models such as IaaS, PaaS and SaaS and the IoT combine and this integration results in the Cloud IoT. The challenges faced by cloud IoT like delay delivery, incorporation of traditional techniques in cloud etc. are solved by integration of software-defined network with cloud IoT (SDN-IoT). This integration is helpful as it provides unified view, agility, and automation.

2   SDN (Software-Defined Network)

Software-defined network, generally abbreviated as SDN, is an excellent approach to manage the network by

a. Making network dynamic
b. Improving performance and monitoring of network by programming network configuration
c. Making network perform more like cloud computing

The static nature of network is dealt with SDN to provide more flexibility to network, which in turn helps in trouble shooting. The main target of SDN is centralization of network intelligence. This aim is achieved by bifurcating the process of forwarding data plane and control plan. Data plane means network packets while the control plane stands for routing process. The routing process have controllers where the complete intelligence of the network is stored. These controllers are termed as SDN controllers. Initially, SDN used OpenFlow protocol according to which remote communication with the network is done using network controllers to decide which path network packets will use for traversing over network switches. With time, this OpenFlow protocol became outdated due to security issues such as denial of service, man-in-middle attack [7], cover communication etc., and Open network Environment and Network Visualization platform came into picture.

2.1   History of SDN Principles

The concept of SDN started from 2004 when IETF proposed ForCES but due to some issues related to increased application further amendments were required and open- source software came into picture. The open-source software was also having security issues so SDNs are combined with NFV or DPI [8]. The history of SDN in

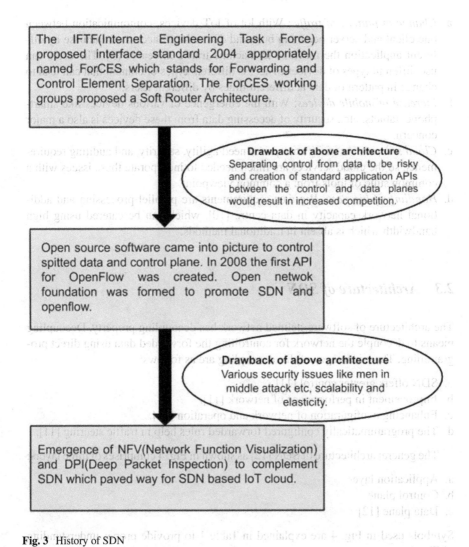

The IFTF(Internet Engineering Task Force) proposed interface standard 2004 appropriately named ForCES which stands for Forwarding and Control Element Separation. The ForCES working groups proposed a Soft Router Architecture.

Drawback of above architecture
Separating control from data to be risky and creation of standard application APIs between the control and data planes would result in increased competition.

Open source software came into picture to control spitted data and control plane. In 2008 the first API for OpenFlow was created. Open network foundation was formed to promote SDN and openflow.

Drawback of above architecture
Various security issues like men in middle attack etc, scalability and elasticity.

Emergence of NFV(Network Function Visualization) and DPI(Deep Packet Inspection) to complement SDN which paved way for SDN based IoT cloud.

Fig. 3 History of SDN

form of its principles can be easily explained using the following flowchart depicted in Fig. 3.

2.2 Need of SDN

With the emergence of mobile and IoT device with cloud computing, the traditional system failed to give impressive results. The main reason that leads to shift of palindrome from traditional methods to new ways are:

a. *Change in pattern of traffic*: With lot of IoT devices, communication between one client and server occurs in bulk and due to lot of access of database by different application the machine-to-machine traffic increases [9]. The users can use different types of devices to access different types of content that can lead to change in pattern of data at different times by different users.

b. *Increase of mobile devices*: With the emergence of various devices like smartphone, tablets, etc., security of accessing data from these devices is also a major concern.

c. *Cloud services*: As many enterprises need agility, security, and auditing requirements, so the cloud services are much needed to incorporate these issues with a common suite of tools from a common viewpoint.

d. *Handling of big data*: Big data requirements are parallel processing and additional network capacity in data centre [10], which can be catered using high bandwidth which is absent in traditional methods.

2.3   Architecture of SDN

The architecture of software-defined network has decoupling property. Decoupling means to decouple the network for controlling the forwarded data using direct programming. The advantages of this decoupling are as follows:

a. SDN offers greater control [11].
b. Improvement in performance of network [11].
c. Enhancing configuration of network and operation [11].
d. The programmatically configured forwarded rules help in traffic steering [11].

The general architecture of SDN [12] as shown in Fig. 4 comprises of three layers:

a. Application layer
b. Control plane
c. Data plane [12]

Symbols used in Fig. 4 are explained in Table 1 to provide proper understanding of Fig. 4.

a. *Application layer*: This layer mainly consists of application software which communicates with control layer; thus, main focus is on network services.

b. *Control layer*: It is a fundamental layer of the software-defined network. It contains SDN controller whose function is to control request from application layer along with managing network devices using the standard protocol [13].

c. *Data plane layer*: It contains physical switch, packet switch and network devices which support interfaces.

d. *Interfaces:* Northbound APIs combined with SDN controller enable various control mechanisms for SDN networks. The southbound API helps applications to control the forwarding devices by flexible programming. The communication

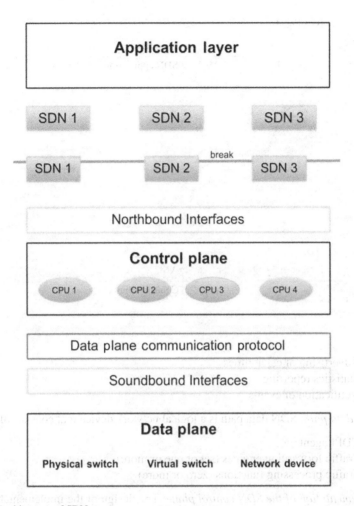

Fig. 4 Architecture of SDN

interface between controller layer and devices of data plane is constructed by southbound API [13].

e. *SDN controller*: It is an extensible software that provides a framework for different user communication to communicate with controller to allow automatic configuration of device in a network. It is a logically centralized entity composed of:

    i. Northbound agent [14]
   ii. SDN control logic [14]
  iii. Control to data plane interface driver (CDPI) [14]

f. *CPDI*: An interface between SDN controller and SDN data paths. Various function of CDPI is to provide:

    i. Programmatic control of all forwarding operations

Table 1 Symbols of Fig. 4

Symbol	Meaning
	SDN application

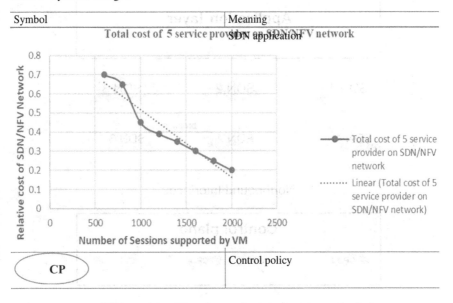

| | Control policy |
| CP | |

   ii. Advertising of capabilities

  iii. Statistics reporting

  iv. Notification of events

g. *SDN data path*: SDN data path is a logical network device that consists of

   i. CDPI agent

  ii. Traffic forwarding engines (set of one or more)

  iii. Traffic processing functions (zero or more)

h. *Implementation of the SDN control plane*: The design of the implementation of SDN control plane is of three types:

   i. Centralized design

  ii. Hierarchical design

  iii. Distributed design

Initially, the focus was on centralized design that simplifies implementation of control logic but due to the limitations in size and increase in dynamics of network hierarchical and distributed approach came into picture. In hierarchical approach, the network-inclusive knowledge is considered by root controllers which are logically centralized and distributed controllers working on a partitioned network view. In distributed approach, controllers work on their local view, and to enhance their knowledge, they exchange synchronization messages. Distributed solutions are more suitable for SDN applications.

i. *SDN flow forwarding*: It uses Ternary Content Addressable Memory (TCAM) also known as associative memory. There can be implementation of software flow table (using open v switch) or hardware (using application-specific integrated circuit) flow table. The matching is done with TCMA, and when no matching is found in TCMA, then request is sent to controller for further instruction. This situation is handed over in three ways:

- Reactive mode: In this mode, the controller takes the request and creates and installs new rules for the corresponding packet in flow table [15].
- Proactive mode: In this mode, the controller finds all possible entries for all possible traffic matches possible for the switch beforehand and fills flow table with it.
- Hybrid mode: In the hybrid mode, there is:

  Reactive mode for a set of traffic due to its flexibility.

  Proactive modes, that is, low-latency forwarding is used for the rest of the traffic.

2.4 Implementation of Issues Related to Security Using Software-Defined Network

By using the property of central view of data used by the controller and its ability to reprogram data in the data plane, many security issues are taken care of; still few concerns of security are a matter of research. Major security concerns taken care of using SDN paradigm are as follows:

a. *Distributed Denial Of service, that is, DDoS, Botnet, and worm propagation*
   The approach for dealing with this problem is three-step process [16]:

   i. Step 1: Using OpenFlow, there is periodic collection of network statistics from the forwarding plane of the network.
   ii. Step 2: Application of classification algorithms on statistics collected in the step 1, to detect any kind of anomalies.
   iii. Step 3: If there is any anomaly detected, the application gives instructions to the controller for reprogramming the data plane for its migration.

b. *Application of MTD algorithm for dealing with some security issues*
   Moving Target Defence algorithm (MTD) [17] is difficult to implement in traditional network (due to lack of central authority to decide protection of every part of system), but in case of SDN there is central controller which is needed to decide which property is to be hid or changed.

c. *Additional security value gained by SDN network using FlowVisor and FlowChecker*
   FlowVisor is used to implement single hardware forwarding plane (for production and development purposes) and multiple separated logical network (for

separating monitoring, configuration and Internet traffic). Each logical topology will be called as slice. FlowChecker checks the new OpenFlow rules that can be used by users to create their own slice [18].

To deploy various security concerns without losing scalability, SDN networks are often combined with different technologies; one popular network is combination of SDN with NFV (Network Function Visualization).

2.5 NFV (Network Function Virtualization)

In NFV architecture, VMs are created on the physical hardware with the help of hypervisor [19]. This is referred to as Virtual Infrastructure. With the help of Open standard API, virtual hardware can be accessed. Using this set of Open Standard APIs, one can create the VNF (Virtual Network Function) with the help of high-level programming languages.

Figure 5 shows the NFV architecture which is composed of two planes:

a. *Management plane*: It comprises of various virtual network functions, that is, VNF.
b. *Data and control plane*: It comprises of virtual machines (VM).

With the help of above planes, the NFV architecture is created whose main function is:

a. Controlling and maintenance of firewall and virtual routing
b. Monitoring and controlling of traffic

VNF architecture is composed of following components:

a. *Physical hardware*: This hardware layer is a bare metal machine and is used for hosting the resources like CPU, storage, memory, I/O, etc.

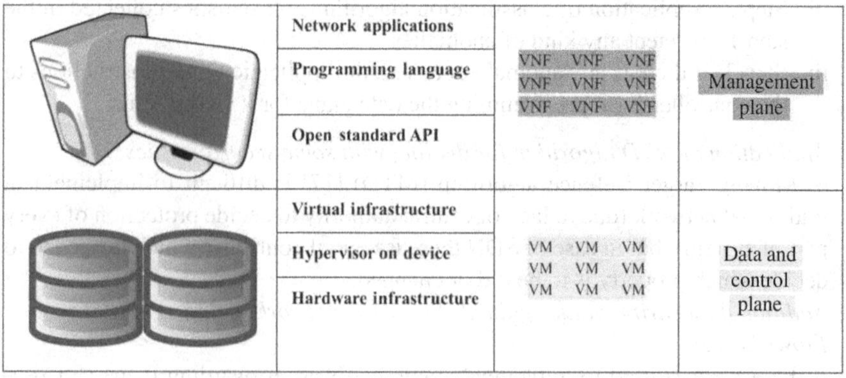

Fig. 5 NFV architecture

b. *Virtual hypervisor layer*: This is the virtual software layer that runs on top of the physical hardware component. Hypervisor is responsible for managing all the underlying resources [20].
c. *Virtual machine*: The software performs approximately all the functionalities of physical platform and depicts the similar architecture as that of physical hardware. A virtual machine uses only small fraction of physical hardware. In this way, one physical hardware can host multiple number of VMs. The number of virtual machines hosted by a bare metal depends on the physical hardware capability and resources' requirements by each VM. Figure 6 explains the realization of VNF in software.

The management of resource allocation and de-allocation for VNF can be done with the help of any software controller like SDN controller in SDN. With SDN controller, VNF-enabled devices can also be controlled, thereby raising the potential of using both the kinds of architecture simultaneously [21].

The usage of SDN and NFV architecture benefits from using the same general COTS server for performing computation of a wide range of applications ranging from computation performed in cellular devices to computations performed in large applications like that of big data handling [22].

2.6   SDN's Relationship with NFV

NFV (Network Function Visualization) complements software-defined network (SDN); it is not dependent on SDN.

NFV uses VNF (virtualized network function). VFNs are software services that run in NFV environment. SDN by using SDN controller gives agility to routers, switches etc. (generic forwarding devices) [23], while NFV gives agility to network

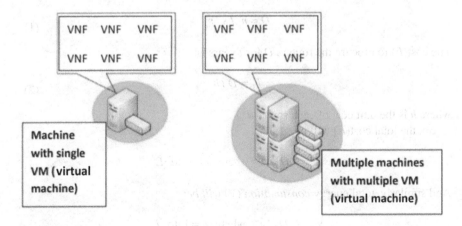

Fig. 6 Realization of VNF in software

Table 2 Difference between SDN and NFV

SDN	NFV
The concept of SDN is used to achieve better programmable and controlled network to get better connectivity	NFV is a concept used for implementing network functions in software
SDN decouples control plan and data plane forwarding controlled by SDN controller	NFV decouples network functions with hardware for agility

application by the virtualized server. The two major differences between NFV and SDN are summarized in Table 2.

Despite the differences between SDN and NFV, they are closely related as SDN provides programming-enabled network connectivity to VNFs for traffic steering, and NFV helps SDN controller to run over cloud, thus providing dynamic migration of controller [23].

2.7 Benefits of SDN-NFV Architecture on Cost and Energy Parameter

Assume there is 5G router d which is capable of handling "n" sessions. Now each service provider s spends h dollars to procure and configure its network. Let us say, the router has its shelf life of t years. This means after t years, it needs to be replaced. Now there are multiple service providers say L. So, in each of the m service providers, there will be a certain central router D_s and particular number of customers, say m. This means that each customer r, on an average, uses n_r sessions.

Let us say that on an average, each router consumes v energy.

Each service provider s requires total D routers to support all its customers, that is,

$$D = n_r / s ? r_m \tag{1}$$

The cost(F) to procure the routers D for t years is:

$$F_1 = D ? h \tag{2}$$

where h is the unit cost of each router d.

So, the total cost(F) will be:

$$F = ? D_s ? h_s \quad \text{where} \quad s = 1 \text{ to } L \tag{3}$$

And similarly, total energy *consumption (V) will be*:

$$V = ? \ D_s ? v_s, \quad \text{where} \quad s = 1 \text{ to } L \tag{4}$$

Let us say that the SDN/NFV architecture provides VMs.

Each VM can handle p sessions in parallel manner, and the cost to lease in a VM is j dollars and similarly energy consumption of a single VM for a day is say w.

To support all the customers, total number of VMs (z) required by each service provider s is:

(Assuming all other input factor to be same)

$$z = n_r / p ? r_m \qquad (5)$$

So, the total cost(J) for t years by service provider s is:

$$J = z ? j ? t \qquad (6)$$

So, the total energy consumption (W) will be:

$$W = z ? w \qquad (7)$$

Assume: $k = \varphi \times p$

$$j ? t = ?? h \qquad (8)$$

$$w = ?? v \qquad (9)$$

From above equations, it can be summarized as:

$$J = ?/ ?? F \qquad (10)$$

So, the total cost for all m service providers:

$$J = F ? (? D_s / ?_s) ? (? D_s / ?_s) \quad \text{where} \quad s = 1 \text{ to } L \qquad (11)$$

$$W = ?/ ?? V$$

Total energy consumption is:

$$W = V ? (? D_s / ?_s) ? (? D_s / ?_s) \qquad (12)$$

where $s = 1$ to L

From Equation 3, 4, 11 and 12, it can be deduced that for SDN-enabled cloud IoT, total cost incurred and energy consumption (L.H.S of 11, 12) is the multiplicative factor of the total cost incurred and energy consumed for cloud IoT without SDN (RHS of 11, 12). These multiplicative factors involve the division by total number of service providers and sessions. So, the multiplicative factors reduce the result of RHS, thereby reducing costs and energy consumed.

With the above analysis and mathematical equations, it can be understood that the usage of SDN/NFV incurs relatively less cost for network when compared to the usage of typical 5G router. The cost reduction is high specifically when VM is powerful enough to handle a large number of sessions. In such cases, the cost of network is comparatively very low. The cost incurred on the 5G router is very high. However, in contrast, VMs created using SDN/NFV can be shared across multiple vendors which in turn further reduce the network cost. There is a substantial amount of cost and energy savings that multiple industries have witnessed using SDN/NFV architecture [24].

Graph 1 depicts how the cost of five different service providers decreases with increase in number of VMs in the network [24].

The above graph concludes that with the multiple service providers (in this case 5), with increase in the number of sessions being handled by VMs through SDN (i.e., X axis), the utilization of VM increases. However, the cost incurred is relatively slow and is in decreasing nature (i.e., Y axis). If the same scenario is considered with the physical hardware devices without SDN, then cost incurred and energy consumption would have been too high because either the hardware's capacity needs to be increased or new additional hardware devices or servers have to be configured. The latter option will also require changes in network configuration, which in turn adds up the cost. The same analysis has been mathematically shown with the help of Equation 11 and 12.

Thus, it is obvious that one hardware device with SDN enabled can host multiple VMs where each VM can in turn host multiple VMs. To conclude, it can be said that one device can handle comparatively more sessions as compared to the situation where a device without SDN can handle fewer sessions.

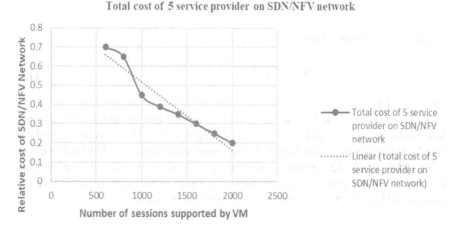

Graph 1: Relationship between cost of SDN/NFV and number of sessions supported by VM

3   Cloud-Based Internet of Things (Cloud IoT)

Cloud computing and IoT are closely related. Cloud computing is basically used for big data analysis and analytics. The various deployment models of cloud computing are explained in Fig. 7:

a. *Private Cloud*: It is exclusively used by single organization.
b. *Public Cloud*: It is open to use by public, for example, Amazon web services.
c. *Hybrid Cloud*: It combines two or more private or public clouds to enable portability of data and applications [25].

Various service models offered by Cloud computing are:

a. *SaaS (Software as a Service)*: In SaaS we can just have an Internet connection and you can use the services without installing and running software applications [26].
b. *PaaS (Platform as a Service)*: Developers generally use PaaS. It provides hardware and software tools over the network [26].
c. *IaaS (Infrastructure as a Service)*: IaaS are very flexible and replaceable networks. With IaaS, the desirable software can be purchased over Internet, and in future it can be upgraded or replaced [26].

The term IoT was first given by Kelvin Ashton in 1999, but he called it as "Internet for things" [27]. IoT means physical objects using sensors [26], software etc. to exchange and process information to other devices using the Internet. The IoT devices are categorized in three categories on basis of their field of usage:

a. Consumer application (smart homes, elder care, etc.)
b. Industrial application (manufacturing, agriculture, etc.)
c. Organizational application (healthcare, transportation, etc.)

When Cloud computing and IoT devices are combined, it is known as Cloud IoT. Working methodology is explained in Fig. 8.
    The major characteristics of cloud IoT are as follows:

a. Cloud-based IoT services is used when you need it, that is, it is on device service.
b. It provides larger connectivity options and hence it borders the network.

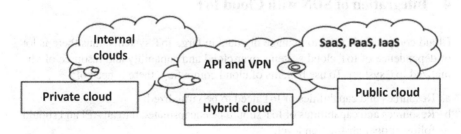

Fig. 7 Types of cloud

Fig. 8 Cloud IoT working

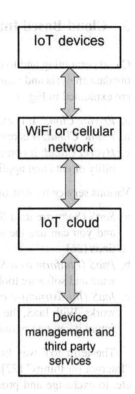

c. In IoT, an IP address is assigned to every IoT device, and in cloud IoT this address can be shared in protected cloud protocol for further analysis.
d. Cloud computing helps IoT by providing better computation and storage.
e. Cloud IoT has properties [28] of:

   • Dynamicity
   • Heterogeneity
   • Geographical distribution of IoT cloud
   • Sheer scale of IoT cloud

4 Integration of SDN with Cloud IoT

Cloud computing is used to manage big data in large IoT systems; thus there is lot of dependence of IoT cloud systems over cloud and capability and resource of virtualized IoT system. To use benefits of cloud computing, there is need of:

a. Resources and capabilities of IoT need to be virtualized.
b. Resources and capabilities of IoT should be encapsulated into an API and should follow proper abstraction level.
c. Automation of resources and capabilities of IoT devices.

The various challenges faced by cloud IoT are:

a. The IoT cloud service mostly relies on the physical infrastructure of IoT, but as IoT resources are provided as roughly grained rigid package, it does not support flexibility, which delays delivery (utility oriented) and reliable consumption of IoT devices [29].
b. The principle of elasticity states to provide required resources dynamically at the time of need. This is done by adjusting the amount of resources of current needs according to changing load pattern. It helps to minimize overprovisioning of load. This elasticity factor is difficult to incorporate elasticity pattern.
c. Cloud computing supports large-scale IoT infrastructure, but the problem is building large-scale dependable and aggregable cloud IoT infrastructure, because such systems lead to convergence of cloud, network, and embedded system. This leads to creation of big hyper-distributed systems which have same security concerns as of traditional methods.
d. The traditional techniques are not feasible to incorporate in cloud IoT infrastructure.

To solve above problems, there is a need of novel model and techniques. One such technique is integration of SDN with IoT cloud.

4.1 Software-Defined IoT Principles

Software-defined IoT (SDN-IoT) comprises resources in the cloud. Such resources are programmed and controlled at run time also. A common example of IoT resources, runtime environment and capabilities are as follows:

a. *IoT resources*: Sensory data stream, etc.
b. *Runtime environment*: Gateways, etc.
c. *Capabilities*: Communication protocol, etc.

The combination of IoT resource with runtime environment with proper capabilities is known as software-defined IoT units (SDN- IoT units). In other words, SDN-IoT units are exposed to the API at different levels of hierarchy to encapsulate lower level functions and resources of IoT cloud which in turn provide governance at runtime also [30].

The principles of SDN-IoT system are as follows:

a. Encapsulation of resources and capabilities of IoT in proper API. It provides a uniform unified view of functions along with configuration of IoT cloud.
b. Consumption of resources and capabilities of IoT should easy to access at every level of granularity. This is helpful for incorporation of agility and self-services consumption.

c. The software-defined IoT units should be declared and configured in API which is well-defined and available in similar types of software libraries. Hence, there should be policy-based specification [31].
d. The processes need to be automated to ensure dynamicity on large-scale IoT systems.

4.2   SDN-IoT Units Conceptual Model

The SDN- IoT units encapsulate two aspects of the resources of IoT and make them available for IoT cloud. The two aspects which encapsulated are as follows:

a. *Functional aspects*: communication possibilities, etc.
b. *Non-functional aspects*: elasticity, cost etc.

In the conceptual model, there is dynamic composition as well as interconnection with software-defined IoT units so that IoT capabilities and resources are allocated to applications at runtime also based on usage [32]. The concept of unit prototype is present. The unit prototype is present in cloud with three capabilities:

a. Functional capabilities
b. Provisional capabilities
c. Governance capabilities

The unit prototype can be configured dynamically and can be controlled at runtime as they are using operating system-level virtualization. Figure 9 shows prototype model of SDN-IoT unit.

In this conceptual model, all three levels deal with the different types of challenges:

a. *Provisional API*: Deals with granularity levels and dynamic coupling of virtual topologies for runtime allocation of units as per demand [33].
b. *Governance API*: It helps in dynamic allocation and removal of resources and hence introduce and implement concept of elasticity.
c. *Functional API*: It deals with cost factor functions for considering IoT devices as utility.

4.3   Classification of Units

Based on purpose as well as capabilities, the unit prototypes are classified into three categories:

a. *Atomic software-defined IoT units*: These are the thinnest grained units whose job is to summarize capabilities of an IoT resource. It has two types of capabilities:

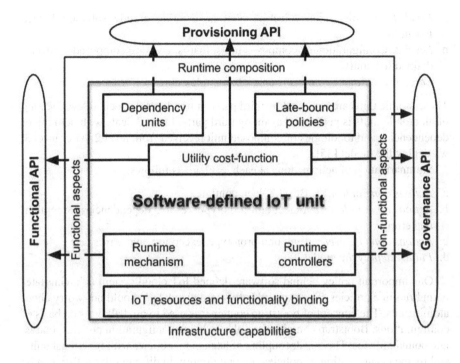

Fig. 9 Unit prototype

- Functional capabilities such as key value storage, in memory image, monitoring, data point controller, gateway runtime, etc.
- Non-functional capabilities such as elasticity, data quality, elasticity, security, and configuration.

b. *Complex software-defined IoT units*: It computes relationship between fine-grained units. It defines internal topology of network by defining an API which integrates the controllers. These controllers help in dynamic reallocation and configuration of network and horizontal elastic scaling of units.

c. *Composed software-defined IoT units*: Just like atomic units, they also have all functional and non-functional capabilities along with additional requirement of mechanism to bind all atomic units, for example, software-defined IoT gateway [34].

4.4 Software-Defined IoT Unit Automatic Composition

Generally, software-defined IoT cloud system's building includes three levels:

a. *Level 1*: Atomic units' selection, that is, creating atomic software-defined IoT units
b. *Level 2*: Configuration of composed units, that is, creating composed software-defined IoT units
c. *Level 3*: Linking of composed unit into complex unit

These atomic units are provided by third party in first level. In second level, adding of all atomic units is performed again by third party. Level 3 deals with defining of dependencies and topologies of composed unit received from level 2. After level 3, we deploy the model [35].

Summarization of actions done at each level are as follows:

a. *Actions done in level 1*: Build, select, configure
b. *Actions done in level 2*: Select unit prototype, resolve dependencies, exception, and errors
c. *Actions done in level 3*: Link unit prototype, exception, and errors
d. *Finally deploy the model*

One important policy behind software-defined IoT cloud system is having late-bound runtime policies. The exposed software-defined API should not worry about atomic units. The concept of bootstrap container is used to bind all units on basis of configurations. Bootstrap container can also redefine configuration policies, that is, late bound policies. There is decoupling of late bound policies with functional units, and by the encapsulation of policies of configuration in different units, they can be easily managed at runtime through management solutions provided by IoT cloud infrastructure in centralized way; the most common example is gateways. The most common approach to making actions of configuration idempotent is to bind them as single units in OS services [35]. Thus, late bound policies and managing protocols for configuration make network flexible and help runtime execution.

5 Benefits of SDN for Cloud IoT

SDN with the help of its interfaces can rapidly configure the cloud IoT devices that generate and transmit data, thereby managing the complexity of network. It also programs the devices and other network elements dynamically according to the requirement. This is done by SDN controller by receiving information from all the network elements. This section discusses the general benefits of using SDN for cloud IoT. Some of its benefits are:

• *Easy network management*

• Cloud IoT has a lot of heterogeneous devices and is of growing nature. This dynamically growing network cannot be handled as well as it cannot be scaled efficiently with traditional technologies. SDN helps in solving the problem by programming the network. SDN's underlying network infrastructure is capable enough to handle such problems with the help of network services [36].

- *Performance improvement*
- SDN can help improve the performance of cloud IoT with the help of its services like load balancing, scheduling bandwidth, and so on by dynamically managing the resources. It can create separate virtual services to separate the traffic and manage it for a specific service provider or any network segment [37].
- *Improvement in security*
- Although this benefit is a debatable area, some industries found that the implementation of SDN improves the security to cloud-based IoT to some extent. This can be achieved with SDN's capability to segment the network and limit the services. Network granular security is complex but can be achieved with the help of virtualization, and most of its functions is carried out by control plane [38].
- *Reduced cost*
- Usage of SDN may reduce the cost of the overall network infrastructure [39]. This benefit has been discussed conceptually and mathematically in detail in Sect. 2.7.
- *Device management*
- Controller can easily manage all the sensors and various other nodes present in the network, and then accordingly apply the appropriate connectivity and can also control them [40].
- *Application management*
- Since SDN allows sharing of resources, SDN can easily customize the network behaviour to support multiple applications [41].
- *Infrastructure management*
- Different network requirements lead to different network policies which are translated with the help of SDN into interfaces to adapt to changing requirements, for example, creation of multi-network environment [42].
- *Energy usage*
- SDN controller itself investigates the computation as well as different other services like scheduling mechanism, network aggregation, and so on. This incurs lower energy consumption [43].
- *Multiple users/owners management*
- SDN supports usage of common infrastructure that can be shared among multiple service providers.
- *Quality of Service (QoS)*
- This benefit is achieved with SDN's help as it can easily translate the network requirements into interfaces and thus improve the performance and QoS, especially for any specific application. Another contributing factor is that the SDN controller has certain commands to control flow scheduling and optimize network usage. With the central controller, cloud QoS management can be easily achieved [44].
- *Mobility and scalability*
- SDN/NFV implementation helps to support scaling and elastic nature of dynamically growing network.

- *CPU use/network expansion*
- Expansion of the network as well as optimization of CPU utilization can be achieved with the help SDN controller as it supports different routing paradigms.
- *Network provisioning in centralized manner*
- With SDN, control of network domain can be viewed in the centralized manner. It abstracts functionalities at higher level and uses it to improve service delivery.
- *Holistic approach*
- SDN makes it easy for IT administrators to work on the enterprise network for certain applications like grid computing without impacting the physical network. This is helpful in dealing with application provisioning or on demand service provisioning.
- *Guaranteed availability of content*

- SDN can easily shape up the traffic with the help of dynamic routing. Thus, even for intense applications like live streaming, video conferencing, network traffic will be easily managed to ensure great user experience and data availability on time.

6 Limitations of SDN-Enabled Cloud IoT

The idea of SDN-enabled cloud-based IoT is fantasizing and seems to be a global solution to all the problems. However, most of the developments or features are yet just theoretical or are not practically feasible. Although various industries are working in this research area and trying to evolve the feature and experimentally validate it, still there are some challenges that are important to be addressed especially in the practical scenario as the real-time network scenario is far more challenging than it looks [45].

Therefore, despite some advantages of using SDN, some challenges affect the performance. These challenges are listed below:

- *Reliability*

- It is one of the important features required for any software as well as network. Whenever any software or network fails, operations should be carried out without any interruption. Traditionally, for the network failure, traffic is routed towards another router or network. However, in SDN, the central SDN controller is the only central authority responsible for handling the entire network. If the controller fails, the entire network will fail leading to poor reliability [46].
- *Scalability*
- Scalability is the potential of any system or network to handle large amount of work, which is likely to grow exponentially in the future. Theoretically SDN is scalable; however, practically it is a very tough task as SDN has logically a centralized controller. SDN controller should be capable enough to support a large number of switches and scalable according to the growing number of IoT devices

or change in network. This is a difficult task because in SDN both the data planes and control planes are decoupled. This means they can grow independently as API is used to connect them. The network changes are mainly effective in the work of control plane [47]. Because of having a standard API for both the planes, the number of switches and nodes increases, causing bottleneck in the network due to the SDN controller.

- *Lower-Level Interface*
- The task of SDN is to translate the network policies to lower-level configuration switches. These switches are coordinated with the help of interface. With the change in network behaviour, policies are required to frequently change, which again might be a tedious task [48]. And frequent translations cause multiple events of asynchronous nature. This coordination is tough and unnecessary if the task is very simple.
- *Performance*
- SDN standards are not compatible with traditional networking. SDN primarily focuses on the flow of network traffic. It will try to measure the performance on two different criteria related to flow, that is, time for flow setup, and the second criterion is the number of flows switched by a controller. Two modes of flow setup are reactive and proactive [49]. This means that these two modes affect the initiation of flow as well as the overheads for the limitation. This demands for further setting for controlling the factors related to SDN, which again is not an easy task.
- *Security*

- Because of the open nature of interfaces, the SDN network is prone to different kinds of network attacks like DDoS and SYN flood attacks. Also, it comprises different security features like integrity, confidentiality, remote access management, timely detection of network threats, and so on.

- The security issue is also important and challenging due to heterogeneous devices present in Cloud IoT, and huge of data is communicated [50].

- Although all the security challenges are the same for wired and wireless SDN, however, usage of wireless medium increases the probability of security issues.

Main security issues are

- SDN forwarding device attack

- Because of the distribution of network traffic by the switches, network failure can occur by the launch of DoS (Denial of Service) attack.
- Control plane threats
- SDN central controller is the vulnerable point whose failure can easily downgrade the entire network.
- Communication channel vulnerability
- Southbound APIs can use different protocols for providing security between the communication of both data channel and control channel. However, the

administrator might be disabled, which suggests another kind of attack, that is, man in the middle attack.

- Fake traffic flows
- Any intruder or buggy device can harm the traffic flow and can also change the direction in which the traffic is moving. This can result in faulty interpretation.
- Authenticity
- An intruder can potentially violate the authenticity of the genuine forwarding devices in the network. This security problem is same as that of traditional network.
- Confidentiality
- Unauthorized user can investigate the information or data potentially by eavesdropping techniques as cloud IoT involves transmission of huge amount of data or even by gaining unauthorized access to SDN devices.
- Availability
- This feature can be compromised if too many security features are implemented. Thus, if all the other security concerns are handled in a peculiar manner, then there is a high chance that data availability or service to authorized user at the right time might be hampered [51].
- Open Programmable API
- Since SDN employs open interfaces and open standard APIs, there is considerable potential for attacks to devices and SDN controllers or even the layers of SDN [52].
- Man-in-the-middle-monitors
- Man-in-the-middle-monitors can basically illegally gain access to information because controllers are not directly connected to switches. Thus, information can be stolen during transmission. This is also known as black hole attack.
- SDN stacks

- Attacks like DDoS, or control flow saturation attack can arise if the transmission of information between control layer and infrastructure layer increases [53].

Let us try to look above security issues with respect to different layers present in SDN:

- Application layer attacks: This involves different rules due to different network domain and the changing network behaviour. For example, injecting malicious code [54].
- Control layer attacks: This involves attacks at controller level, channel level as well as at connection between controllers and switchers, for example, DoS attacks, gaining unauthorized access.
- Infrastructure layer attacks: This involves attacks like man-in-the-middle attack or DoS attack since information is being transmitted between different layers [55].

- *Controller placement*

- In Cloud, special attention must be paid to the position of controller and its configuration according to topology and network to maximize the reliability [56]. Failure of the controller must be avoided and if happened due to unforeseen circumstances, it should be handled by any other controller. The controller needs to be configured in a way to handle multiple network paths. This is also important and challenging keeping in mind the large number of IoT devices, and the huge amount of data is being generated by them and that too of heterogeneous nature [57, 58].

There are some other challenges that should also be addressed, such as:

a. SDN has routing table of limited size for switches.
b. Implementation of cloud IoT further adds up new environment changes and, therefore, challenges for SDN. This requires further enhancement of SDN features and its configuration. Initially, SDN is not responsible for gathering data. It only deals with the flow of traffic. However, because of integration with Cloud IoT, it needs to investigate data issues as well. It is required to deal with this heterogeneity and scalability, which is not an easy task.
c. For integrating SDN with Cloud IoT, special interface is required that need to be flexible yet efficient.
d. An efficient way to calculate the influence of network traffic generated by cloud IoT on SDN is needed to overcome any issues.
e. SDN has a big challenge to handle the mobility of devices and handle different protocols for the same.
f. Another challenge that SDN must deal with is interface. Initially, it deals with Southbound and Northbound APIs. However, integration with cloud IoT requires creating device-specific interfaces that are very difficult in the ever-changing network.

7 Conclusion and Future Scope

The growing network challenges because of emerging technologies like IoT and Cloud Computing and how these issues can be solved with the virtualization technologies like SDN/NFV were discussed. Then these virtualization technologies, that is, SDN, NFV were explained in detail along with the concept of their integration as well integrating SDN with cloud-based IoT. Later, the chapter briefly overviewed the problems of cloud IoT and how it was solved by introducing SDN into it. Furthermore, the benefits of using SDN were studied in detail, which was comprised of two sections. The first section described the comparison of cost incurred and energy consumption in terms of mathematical equations for both the scenarios with/without SDN. The second section investigated general benefits that are introduced by SDN into the cloud IoT network.

The chapter also discussed the challenges being possessed by SDN over cloud IoT which needs to be addressed, and the challenges being faced because of implementation of SDN in cloud-based IoT network. Also, there is a need to investigate the existing methods and develop algorithms to cater to the changing network's needs. This area's scope is to develop the application-specific algorithms that should handle the growing amount of heterogeneous network. Focus should also be given on the parameters impacting the performance. Due to integration of SDN and cloud IoT, there might be some new challenges specifically related to network management and maintenance that needs to be studied and addressed carefully. Some of the major challenges of SDN like providing QoS, protection against security attacks should be handled in future in a better way yet in a cost-effective manner. At last, there is always a scope of improvement in costs incurred, energy consumption and various other important parameters advocating for SDN usage in cloud-based IoT.

Further, the work can be carried with the improvement of APIs involved as APIs play significant role in SDN. Instead of having multiple APIs, one standard API can be developed. The connectivity between the layers can be improved and made secure to avoid any potential performance problems and attacks.

Another dimension in which the future work can be carried on is Edge IoT. Since Cloud computing's movement to Edge/Fog computing is being initiated, the IoT network can also be translated into Fog. SDN can be enabled on edge nodes with the help of containers and can be clubbed with cloud and other network elements. This integration can further be used to handle the problems of distributed processing occurring at the network. By having edge computing with IoT and SDN, the network traffic can be further reduced as the edge IoT devices will not just generate the data but will also process it then and there only. Only the necessary data that needs to be stored at some other server will be transmitted to cloud thereby shedding off the load of the cloud. This technique will reduce the network traffic to a considerable amount. With the help of SDN on edge nodes, the required services can be provisioned. The same device can perform different roles at the network, thereby further reducing the cost for procuring several computing edge devices.

References

1. Manfredi, S. (2014). Congestion control for differentiated healthcare service delivery in emerging heterogeneous wireless body area networks. *IEEE Wireless Communications, 21*(2), 81–90.
2. Shirmarz, A., & Ghaffari, A. (2020). Performance issues and solutions in SDN-based data center: A survey. *The Journal of Supercomputing, 76*, 7545–7593.
3. Lian, R., Shih, T. Y., Yin, Y., & Behdad, N. (2017). A high-isolation, ultra-wideband simultaneous transmit and receive antenna with monopole-like radiation characteristics. *IEEE Transactions on Antennas and Propagation, 66*(2), 1002–1007.
4. Midha, S., & Tripathi, K. (2021). Extended security in heterogeneous distributed SDN architecture. In G. Hura, A. Singh, & L. Siong Hoe (Eds.), *Advances in communication and computational technology* (Lecture notes in electrical engineering) (Vol. 668). Springer.

5. Khondoker, R. (Ed.). (2018). www.springer.com). *SDN and NFV security: Security analysis of software-defined networking and network function virtualization.* Springer.
6. Saha, D., Shojaee, M., Baddeley, M., & Haque, I. (2020, June). An energy-aware SDN/NFV architecture for the internet of things. In *2020 IFIP networking conference (networking)* (pp. 604–608). IEEE.
7. Salim, M. M., Rathore, S., & Park, J. H. (2020). Distributed denial of service attacks and its defenses in IoT: A survey. *The Journal of Supercomputing, 76,* 5320–5363.
8. Raj, J. S., & Smys, S. (2019). Virtual structure for sustainable wireless networks in cloud services and enterprise information system. *Journal of ISMAC, 1*(03), 188–205.
9. Tur, M. R., & Bayindir, R. (2020). The requirements of the technique of communication from machine to machine applied in smart grids. In *Artificial intelligence and evolutionary computations in engineering systems* (pp. 405–418). Springer.
10. Wang, X., Yang, L. T., Kuang, L., Liu, X., Zhang, Q., & Deen, M. J. (2019). A tensor-based big-data-driven routing recommendation approach for heterogeneous networks. *IEEE Network, 33*(1), 64–69.
11. Montazerolghaem, A. (2021). Software-defined load-balanced data center: Design, implementation, and performance analysis. *Cluster Computing, 24,* 591–610.
12. Bhattacharjya, A., Zhong, X., Wang, J., & Li, X. (2020). CoAP—application layer connection-less lightweight protocol for the Internet of Things (IoT) and CoAP-IPSEC security with DTLS supporting CoAP. In *Digital twin technologies and smart cities* (pp. 151–175). Springer.
13. Ahmad, S., & Mir, A. H. (2020). Scalability, consistency, reliability and security in SDN controllers: A survey of diverse SDN controllers. *Journal of Network and Systems Management, 29*(1), 1–59.
14. Li, S., & Hu, Q. (2020). Dynamic resource optimization allocation for 5G network slices under multiple scenarios. In *2020 Chinese control and decision conference (CCDC)* (pp. 1420–1425). IEEE.
15. Rzepka, M., Borylo, P., Lason, A., & Szymanski, A. (2020). PARD: Hybrid proactive and reactive method eliminating flow setup latency in SDN. *Journal of Network and Systems Management, 28*(4), 1547–1574.
16. Özçelik, İ., & Brooks, R. (2020). *Distributed denial of service attacks: Real-world detection and mitigation.* CRC Press.
17. Liu, B., & Wu, H. (2020). Optimal D-FACTS placement in moving target defense against false data injection attacks. *IEEE Transactions on Smart Grid, 11,* 4345–4357.
18. Kaur, K., Mangat, V., & Kumar, K. (2020). Architectural framework, research issues and challenges of network function virtualization. In *2020 8th international conference on reliability, Infocom technologies and optimization (trends and future directions)(ICRITO)* (pp. 474–478). IEEE.
19. Kaur, K., Mangat, V., & Kumar, K. (2020). A comprehensive survey of service function chain provisioning approaches in SDN and NFV architecture. *Computer Science Review, 38,* 100298.
20. Rista, A., Ajdari, J., & Zenuni, X. (2020). Cloud computing virtualization: A comprehensive survey. In *2020 43rd international convention on information, communication and electronic technology (MIPRO)* (pp. 462–472). IEEE.
21. Das, T., Sridharan, V., & Gurusamy, M. (2019). A survey on controller placement in sdn. *IEEE Communication Surveys and Tutorials, 22*(1), 472–503.
22. Zhang, L., Li, C., Wang, P., Liu, Y., Hu, Y., Chen, Q., & Guo, M. (2019). Characterizing and orchestrating NFV-ready servers for efficient edge data processing. In *Proceedings of the international symposium on quality of service* (pp. 1–10). IEEE.
23. Haggag, A. (2019). Network optimization for improved performance and speed for SDN and security analysis of SDN vulnerabilities. *International Journal of Computer Networks and Communications Security, 7*(5), 83–90.
24. Ferrús, R., Koumaras, H., Sallent, O., Agapiou, G., Rasheed, T., Kourtis, M. A.,Ahmed, T. (2016). SDN/NFV-enabled satellite communications networks: Opportunities, scenarios and challenges. *Physical Communication, 18,* 95–112.

25. Gopalakrishnan, B., & Maheswari, U. (2019). Research on enterprise public and private cloud service. *International Journal of Innovative Technology and Exploring Engineering (IJITEE).* ISSN: 2278–3075, *8*(6S4), 1453–1459.
26. Fehér, D. J., & Sándor, B. (2019, May). Cloud SaaS security issues and challenges. In *2019 IEEE 13th international symposium on applied computational intelligence and informatics (SACI)* (pp. 000131–000134). IEEE.
27. Palmaccio, M., Dicuonzo, G., & Belyaeva, Z. S. (2021). The internet of things and corporate business models: A systematic literature review. *Journal of Business Research, 131*, 610–618.
28. De Donno, M., Tange, K., & Dragoni, N. (2019). Foundations and evolution of modern computing paradigms: Cloud, iot, edge, and fog. *IEEE Access, 7*, 150936–150948.
29. Li, W., Liao, K., He, Q., & Xia, Y. (2019). Performance-aware cost-effective resource provisioning for future grid IoT-cloud system. *Journal of Energy Engineering, 145*(5), 04019016.
30. Nikolov, N., & Nakov, O. (2019). Creating architecture and software of embedded systems with constrained resources and their communication to the IoT cloud. In *2019 X national conference with international participation (ELECTRONICA)* (pp. 1–4). IEEE.
31. Yadav, S. P., & Yadav, S. (2020). Image fusion using hybrid methods in multimodality medical images. *Medical & Biological Engineering & Computing, 58*, 669–687. https://doi.org/10.1007/s11517-020-02136-6
32. Nogueira, L., Barros, A., Zubia, C., Faura, D., Gracia Pérez, D., & Miguel Pinho, L. (2020). Non-functional requirements in the ELASTIC architecture. *ACM SIGAda Ada Letters, 40*(1), 85–90.
33. Durrani, M., Nawaz, S. N., Chinthalapati, E., Zhang, Y., & Payankulath, S. (2015). U.S. Patent Application No. 14/721,978.
34. Jararweh, Y., Alsmirat, M., Al-Ayyoub, M., Benkhelifa, E., Darabseh, A., Gupta, B., & Doulat, A. (2017). Software-defined system support for enabling ubiquitous mobile edge computing. *The Computer Journal, 60*(10), 1443–1457.
35. Lemoine, F., Aubonnet, T., & Simoni, N. (2020). IoT composition based on self-controlled services. *Journal of Ambient Intelligence and Humanized Computing, 11*, 5167–5186.
36. Kreutz, D., Ramos, F. M., Verissimo, P. E., Rothenberg, C. E., Azodolmolky, S., & Uhlig, S. (2014). Software-defined networking: A comprehensive survey. *Proceedings of the IEEE, 103*(1), 14–76.
37. Li, X., Casellas, R., Landi, G., de la Oliva, A., Costa-Perez, X., Garcia-Saavedra, A., Vilalta, R. (2017). 5G-crosshaul network slicing: Enabling multi-tenancy in mobile transport networks. *IEEE Communications Magazine, 55*(8), 128–137.
38. Kutscher, D. (2016). It's the network: Towards better security and transport performance in 5G. In *2016 IEEE conference on computer communications workshops (INFOCOM WKSHPS)* (pp. 656–661). IEEE.
39. Omnes, N., Bouillon, M., Fromentoux, G., & Le Grand, O. (2015). A programmable and virtualized network & IT infrastructure for the internet of things: How can NFV & SDN help for facing the upcoming challenges. In *2015 18th international conference on intelligence in next generation networks* (pp. 64–69). IEEE.
40. Velasco, L., Piat, A. C., Gonzlez, O., Lord, A., Napoli, A., Layec, P., Cugini, F. (2019). Monitoring and data analytics for optical networking: Benefits, architectures, and use cases. *IEEE Network, 33*(6), 100–108.
41. Nisar, K., Welch, I., Hassan, R., Sodhro, A. H., & Pirbhulal, S. (2020). A survey on the architecture, application, and security of software defined networking. *Internet of Things, 12*, 100289.
42. Kim, J. A., Park, D. G., & Jeong, J. (2020). Design and performance evaluation of cost-effective function-distributed mobility management scheme for software-defined smart factory networking. *Journal of Ambient Intelligence and Humanized Computing, 11*, 2291–2307.
43. Nour, B., Mastorakis, S., & Mtibaa, A. (2020). Compute-less networking: Perspectives, challenges, and opportunities. *IEEE Network, 34*, 259–265.

44. Yucel, S. (2019). QoS management and traffic engineering for virtual SDN services. In *2019 international conference on computational science and computational intelligence (CSCI)* (pp. 1448–1453). IEEE.
45. Ahad, M. A., Tripathi, G., Zafar, S., & Doja, F. (2020). IoT data management—Security aspects of information linkage in IoT systems. In *Principles of internet of things (IoT) ecosystem: Insight paradigm* (pp. 439–464). Springer.
46. Khan, R., & Amjad, M. (2019). Mutation-based genetic algorithm for efficiency optimisation of unit testing. *International Journal of Advanced Intelligence Paradigms, 12*(3–4), 254–265.
47. Latif, Z., Sharif, K., Li, F., Karim, M. M., Biswas, S., & Wang, Y. (2020). A comprehensive survey of interface protocols for software defined networks. *Journal of Network and Computer Applications, 156*, 102563.
48. Mishra, P., Puthal, D., Tiwary, M., & Mohanty, S. P. (2019). Software defined IoT systems: Properties, state of the art, and future research. *IEEE Wireless Communications, 26*(6), 64–71.
49. Yadav, S. P., Mahato, D. P., & Linh, N. T. D. (2020). *Distributed artificial intelligence: A modern approach* (1st ed.). CRC Press. https://doi.org/10.1201/9781003038467
50. Ari, A. A. A., Ngangmo, O. K., Titouna, C., Thiare, O., Mohamadou, A., & Gueroui, A. M. (2019). Enabling privacy and security in cloud of things: Architecture, applications, security & privacy challenges. *Applied Computing and Informatics.* 1(1)
51. Kumar, M., & Srivastava, S. (2018). Image authentication by assessing manipulations using illumination. *Multimedia Tools and Applications, 78*(9), 12451–11246. https://doi.org/10.1007/s11042-018-6775-x
52. Yadav, S. P. (2020). Vision-based detection, tracking and classification of vehicles. *IEIE Transactions on Smart Processing and Computing, SCOPUS,* ISSN: 2287-5255, 9(6), 427–434. https://doi.org/10.5573/IEIESPC.2020.9.6.427
53. Singh, J., & Behal, S. (2020). Detection and mitigation of DDoS attacks in SDN: A comprehensive review, research challenges and future directions. *Computer Science Review, 37*, 100279.
54. Yaacoub, J. P. A., Salman, O., Noura, H. N., Kaaniche, N., Chehab, A., & Malli, M. (2020). Cyber-physical systems security: Limitations, issues and future trends. *Microprocessors and Microsystems, 77*, 103201.
55. Varshney, T., Sharma, N., Kaushik, I., & Bhushan, B. (2019). Architectural model of security threats & their countermeasures in IoT. In *2019 international conference on computing, communication, and intelligent systems (ICCCIS)* (pp. 424–429). IEEE.
56. Lu, J., Zhang, Z., Hu, T., Yi, P., & Lan, J. (2019). A survey of controller placement problem in software-defined networking. *IEEE Access, 7*, 24290–24307.
57. Tzounis, A., Katsoulas, N., Bartzanas, T., & Kittas, C. (2017). Internet of Things in agriculture, recent advances and future challenges. *Biosystems Engineering, 164*, 31–48.
58. Aggarwal, A., Rani, A., & Kumar, M. (2019). A robust method to authenticate license plates using segmentation and ROI based approach. *Smart and Sustainable Built Environment.* https://doi.org/10.1108/SASBE-07-2019-0083

Malware Discernment Using Machine Learning

Vivek Srivastava and Rohit Sharma

1 Introduction

MALWARE can be characterized as the code developed without the permission of the owner to infiltrate or injure an electronic machine [1]. For all kinds of viruses, malware is basically a non-classified description. Computer file contagiousness and autonomous malware constitute a straightforward categorization of malware. Malware can actually also be defined by its particular type: backdoors, worms, spyware, Trojans, rootkits, adware, and so on. Malware detection [5] by machine learning is difficult because all current malware programs seem to have several layers to escape suspicion or use side strategies to simply turn to a newer version at short periods of time in order to prevent them from being recognized by any antivirus subroutines [2].

2 Malware Definition

Viruses, ransomware, and spyware, including a spectrum of different malicious codes, sound like malware as a blanket term. Tachygraphy typically includes code designed by swindlers for malicious applications, built to cause unnecessary system and data interference or to govern unauthorized network access [3]. Generally, malware is transmitted via email in the form of a reference or file that requires the user to tap on the attachment or open the malware executing file.

V. Srivastava (✉) · R. Sharma
Dr. Ambedkar Institute of Technology for Handicapped,
Kanpur, Uttar Pradesh, India

© The Author(s), under exclusive license to Springer Nature Switzerland AG 2022
F. Al-Turjman et al. (eds.), *Transforming Management with AI, Big-Data, and IoT*, https://doi.org/10.1007/978-3-030-86749-2_12

3   Latest Trends and Attacks

Here, we present the latest trends and attacks with the reference of an antivirus company Quickheal [6]; in their report, we found the current trends of attacks, some of which we are mentioning in this chapter.

3.1   Ransomware Exploring New Techniques for Process Code Injection

Process code injection is a very popular technique among malware authors to evade security products. Process hollowing is an injection technique where the legitimate process is created in suspended mode, its memory is overwritten with malicious code, and the process is resumed. It seems like a legitimate process performs all the malicious activities, so it is untouched by security products. The new ransomware Mailto or Netwalker is using this old trick in a new way. Instead of creating a process in suspended mode, 'Debug Mode' is used. Using the debug API WaitForDebugEvent, it receives the method and thread information. A segment is then developed with a size that is replicated from the specimen and whole file data. It then manually resolves the relocation. The sample contains an encrypted JSON file in the resource section having required information like a key for generating ID, that is, extension to be added to encrypted files, base64 encoded ransom note, whitelisted paths, and email ids which are part of the extension. The ID is created using the key maintained in decrypted JSON under the 'mpk' tag, the machine name retrieved, and the hardware profile information about the machine being infected. SHA-256 is determined from such inputs, and the first five characters of the result are used as the file extension ID. The ransom note file name is also preserved in the same way as the produced ID. Since the 1980s, malware has posed a threat to servers, networks, and infrastructure. While there are two main technologies to combat this, most organisations depend almost entirely on one technique, the decade-old signature-based technique. The more sophisticated method of detecting malware through behaviour analysis proposed and introduced by the authors in [18] is gaining popularity quickly, but it is still relatively unknown.

3.2   Info-Stealer Hidden in the Phishing Emails!

Cybercriminals [22] pry on the data precious to you! The data consists of your system details, name, software installed, browsing history; cookies stored on the disk; and saved passwords. We have observed multiple phishing emails in this quarter with contents which entice the end user to download a malware encapsulated as fake software or a fake update. These software names have strings like demo, free, cracked or plugin, etc. These malware payloads are often either placed on compromised websites or popular file-sharing service platforms. Once the malware is executed, it starts getting computer details using Windows APIs and stores it in a file. Sometimes, malware downloads a few supporting files, which might be useful for retrieving data – we saw a recent case wherein five to six supporting DLLs were downloaded for Mozilla's Firefox browser. For each kind of data, a separate file is maintained with almost all data stored in a SQLite format. All the stolen data is compressed in a single file and sent to the attacker. We have found some variations in the way of sending data. Some malware contained pre-existing CnC details, and in some cases, there were few mail IDs seen where this data was being sent using Microsoft's CDO library over port 465. To remove the traces, malware finally deletes all the stolen data and downloaded DLLs from the victim's system.

4 Types of Malware

4.1 Virus

Viruses tag their malicious code for cleaning code, the most severe malware method, and look for an inexperienced device user or motorized process to execute the corrupt file. They can expand quickly and widely, like a living virus, posing a danger to the elementary features of systems, manipulating files and preventing authentic users from accessing their computers. They are located within a normal executable set of code.

4.2 Worms

Analogously, they infect computer systems like a real worm plaguing the human body. They amalgamate their way throughout the network, connecting to sequential devices starting from one infected computer to begin the transmission of infection. This kind of malware can swiftly infect common communications systems.

4.3 Spyware

It is intended at spying on what an individual is doing. This malicious code, concealed on a computer in the background, steals processed data without the user's knowledge, such as credentials for bank cards, watchwords and other confidential information.

4.4 Trojans

As licit software, this form of malware stashes inside or dissembles itself. Acting discreetly can cleft protection by building backdoors that offer easy access to other malware variants.

4.5 Ransomware

Ransomware is also recognized as spyware and comes at a high cost. Competent of trapping networks and locking out users before a ransom is credited, ransomware has targeted, with exorbitant consequences, some of the giant organizations in the world today.

5 Malware Detection Techniques

Here we have presented the hierarchical structure of malware detection techniques [8, 10] which makes easier to understand the method of malware detection (Fig. 1).

5.1 Signature Based

It is an anti-malware perspective that recognizes the presence of a malware infection or illustration by matching the software in question with at least one-byte code pattern with the set of signature data from established malicious programs, also known as blacklists. This detection scheme is based on the belief that signature-based detection is the most widely used approach for anti-malware systems represented by trends.

5.2 Susceptible to Evasion

These byte patterns are also well documented because the signature byte patterns are fetched from known malware. Therefore, hackers can easily avoid them by using basic garble techniques such as no-ops insertion and re-ordering code. It is, therefore, possible to refit malware code and prevent signature-based detection.

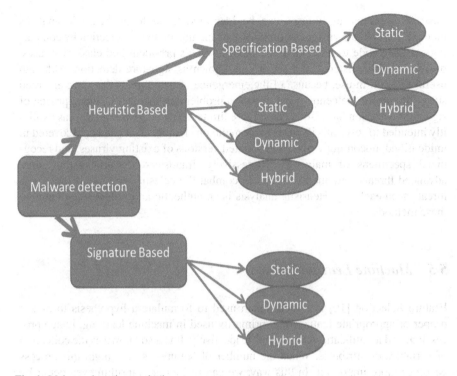

Fig. 1 Malware detection method

5.3 Zero-Day Attacks

Zero-Day Attack is an intrusion that targets a potentially significant vulnerability in software security [19] that might not be identified to suppliers or developers. To minimize the vulnerability to the software user, the software developer should fix the vulnerability as soon as possible after it is found. A software improvement is called the alternative. It is also possible to use zero-day attacks to target the Internet of Things.

5.4 Heuristic Based

Heuristic evaluation emanates from the primitive Greek word meaning 'to discover', and is a process of exploration, learning and problem-solving that uses semantics, assumptions or informed guesses to find an effective response to a specific issue. Although this problem-solving approach may not be efficient, when applied to computer processes where a fast response or timely warning is required based on inherent prudence, it can be highly efficient. By checking code for mistrustful properties, viruses can be identified through the heuristic analysis method. Reasonable virus detection methods include malware detection by comparing software code to the code of recognized type's pre-identified class of viruses, processed and registered in a set of data known as signature detection. Although useful and still in use, because of the emergence of new threats that have emerged since the turn of the century and remain a problem all the period, the approach of signature detection has also become more limited. The heuristic model was explicitly intended to solve this issue, classify malicious features that can be discovered in unidentified, upcoming viruses and updated versions of existing viruses, and recognized specimens of malware. Progressively, fraudsters are propagating more advanced threats; few methods used to combat the celestial volume of these new threats seen each day. Heuristic analysis is an authentic and powerful tool among these methods.

5.5 Machine Learning Based

Feature Selection [10] is a mechanism used to formulate a hypothesis to pick a proper or appropriate feature. It is primarily used in machine learning, image processing, and identification of patterns, respectively. It is also known as the collection of variables or attributes. When the number of features is very high, this process becomes more important. In this way, we can make our algorithm even better by selecting only the significant and important function. With this, our machine's training and evolution time can be decreased. The collection of features is primarily used for few reasons like machine learning prerequisites can be assembled within the deadline; the Machine Learning Algorithm can be trained to reduce the model's complexity such that the model is easier to understand. Feature selection can be accomplished by various methods like filter method, wrapper method and embedded method (Fig. 2).

Feature selection or attribute selection is very basic and essential part of machine learning in order to develop a model for malware detection because without identifying any feature or proper attribute an authentic model cannot be developed.

6 Malware Analysis Techniques

Malware perusal is the task of determining a deeply suspicious file or URL's actions and intent. The perusal performance assists in identifying and alleviating the significant risk. The main advantage of malware analysis helps event responders and security professionals [11]: Idealistically, triage events by severity level *discover* secret compromise indicators that should be blocked, *increase* the effectiveness of compromise warning and warning indicators and *deepen* sense while hunting for threats (Fig. 3).

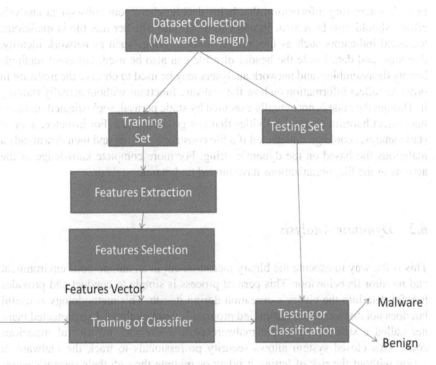

Fig. 2 Machine learning method for identification of malware and benign

Fig. 3 Malware analysis
techniques

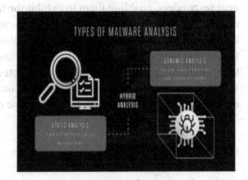

6.1 Static Analysis

Basic static perusal does not really require the literal implementation of the code
[7]. The file is instead investigated by static perusal for signs of malicious intent.
Recognizing malicious systems, libraries or packed files may be valuable. This is
the procedure through which a binary is analysed without conducting it. It is quick
to implement and enables you to retrieve the archive associated with the binary
offender. Static perusal may not reveal all the information required, but it can often

provide interesting information that helps decide where your subsequent analysis efforts should also be aimed. In order to determine whether this file is malicious, technical indicators such as file names, hashes, strings such as network identity, domains, and data inside the header of a file can also be used. Different methods having disassemblers and network analysers may be used to observe the malware in order to collect information on how the malware functions without actually running it. Though the code is not actually executed by static perusal, sophisticated malware may inject harmful runtime activities that can go undetected. For instance, a naive static analysis could go undetected if a file constructs a string and then downloads a malicious file based on the dynamic string. For more complete knowledge of the actions of the file, organizations have moved to dynamic analysis.

6.2 Dynamic Analysis

This is the way to execute the binary measurement in an autonomous environment and monitor its behaviour. This perusal process is simple to conduct and provides helpful data into the binary's operation during its run. This methodology is useful but does not disclose all of the unkind program's functionalities. In a protected manner called a sandbox, dynamic malware perusal executes mistrustful malicious code. This closed system allows security professionals to track the malware in action without the risk of letting it infect or migrate through their system's enterprise network. Dynamic perusal offers deeper visibility to hazard hunters and incident responders, enabling them to exhibit the true nature of a threat. As a secondary advantage, the time it would take to reverse engineer a file to come across the malicious code is eliminated by automatic sandboxing. The dynamic perusal provocation is that adversaries are knowledgeable, and they know there are sandboxes out there, so they have become really good at detecting them. Adversaries conceal code within them, which may remain dormant until certain standards are fulfilled, to mislead a sandbox. Only then, the subroutine operates.

6.2.1 Cuckoo Sandbox

Cuckoo Sandbox [3] is an automated malware perusal framework with boundless application possibilities that are newfangled, highly scalable, and actually completely open source. Through definition, malicious files like executable file written in any source code, normal office documents, any kind of pdf files, emails and also malicious websites can be examined in virtualized environments under different kinds of operating systems like Windows, Linux, MacOS and Android. Trace API calls and normal file activities sublimate this into high-level facts and signatures that everyone can comprehend. Dump and evaluate, even when encrypted with secure socket layers, network traffic (Fig. 4).

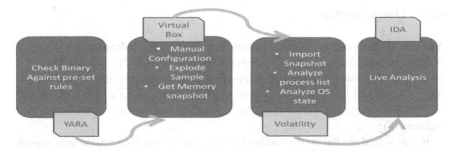

Fig. 4 Cuckoo sandbox analysis

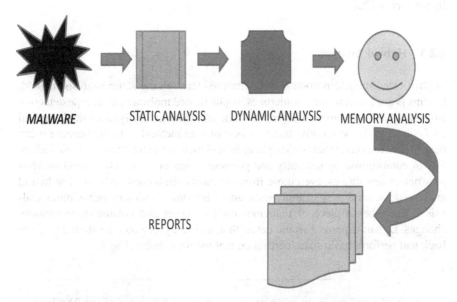

Fig. 5 Limon sandbox analysis

With the help of vernacular network routing support, it can drop all traffic or route it through InetSIM, an access point, or a virtual private network. Perform advanced storage analysis of the infected virtualized system using flickery as well as on the precision of the analyse memory through YARA. Due to the extreme available to all nature of Cuckoo and the large modular architecture, every part of the perusal setting, the processing of research results and the reporting stage can be configured. Cuckoo gives you all the specifications to quickly incorporate the sandbox according to their requirements, like one can use the sandbox in compatible format without fulfilling the licensing requirements. Because of its availability to all features, sandbox is very much used in the current environment to detect malware, and results are also satisfactory.

6.2.2 Limon Sandbox

A Limon sandbox is the most widely used product in malware detection. It basically collects the malware perform static analysis, the result of the static analysis process through dynamic analysis then performs the memory analysis feature final result. Basic process of Limon sandbox is to analyse the malware in a controlled environment, check its actions, and its sub-processes to find out the nature and intent of the malware (Fig. 5).

It tries to find out the process activity of malware, its communication mechanism with files and network connection; it also completes memory perusal and stores the analysed artifacts for later perusal. Such attacks are even possible in recent wireless communication technologies such as cognitive radio sensor networks as proposed by authors in [20].

6.2.3 Hybrid Analysis

Sophisticated malicious code may be detected through the static analysis method, but this is not a secure way; sometimes, sophisticated malware can escape detection from sandbox technology. Merging of static and dynamic analysis methods, hybrid analysis presents the security team the best of both methods – firstly because it can detect malicious code that is trying to hide, and then can extract innumerable indicators of compromise by statically and previously unseen code. The hybrid analysis may detect new threats, even those from the most enlightened malware. The hybrid analysis does, for example, apply static analysis to information by behavioural analysis – like when a piece of malicious code executes and creates some memory changes. Dynamic perusal would detect that, and analysts would be alerted to circle back and perform basic static perusal on that memory dump (Fig. 6).

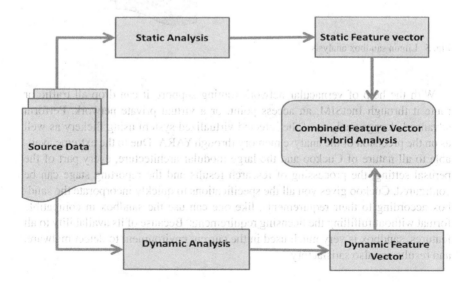

Fig. 6 Hybrid analysis process

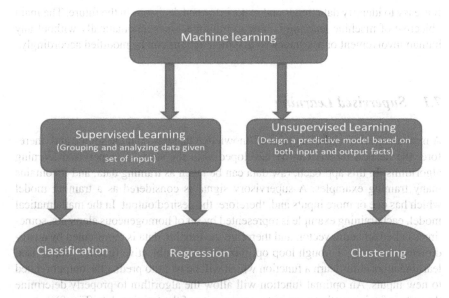

Fig. 7 Machine learning process bifurcation

In hybrid analysis method, the source data is processed with both static and dynamic methods. We get the static feature vector and the dynamic feature vector after processing data from both static and dynamic methods. After merging both feature vectors, we get the combined feature vector, which is also considered as hybrid analysis.

7 Machine Learning

If we think about machine learning, it takes a very technical term to listen. But if you understand about it properly, then it is very easy process which is used in almost all places nowadays. It is basically a research in computer algorithm that moves ahead by the experience. This is a type of learning in which the machine learns a lot of things without programming it explicitly. This is a type of application of AI (artificial intelligence) that gives this ability to the system so that they automatically learn from their experience and improve themselves. This may not sound true, but it is true because nowadays AI has become so advanced that it can make machines do many things which were not possible before. Since machine learning can easily handle multidimensional and multidiverse data in a dynamic environment, there are thousands of advantages of machine learning that we use in our daily work (Fig. 7).

As we have already known, machine learning is a form of artificial intelligence (AI) application that gives programs the ability to learn spontaneously and enhance themselves when needed. They use their own knowledge in order to do this, not because they are programmed directly. Computer programs development of machine learning a focus foretells at could access the data and then use it for his own learning. Learning this starts with data observations, such as direct experience or training.

It is easy to identify data trends and make informed decisions in the future. The main objective of machine learning is how to train machines automatically without any human involvement or assistance so that their actions can be modified accordingly.

7.1 Supervised Learning

A mathematical model of a group of knowledge that has both the source and, therefore, the desired targets can be developed with the study of supervised learning algorithms. In this approach, raw data can be taken as training data, and it contains many training examples. A supervisory signal is considered as a training model which has one or more inputs and, therefore, the desired output. In the mathematical model, each training example is represented by set of homogeneous elements, sometimes called a feature vector, and therefore the training data is represented by a two-dimensional array. Through loop optimization of an objective function, supervised learning algorithms learn a function which will be wont to predict the output related to new inputs. An optimal function will allow the algorithm to properly determine the output for inputs that were not a community of the training data (Fig. 8).

It is believed that an algorithm that improves the accuracy of its outputs or predictions over time has learned to perform that job.

7.1.1 The Multiple Methods of Supervised Learning

Regression

Using training data, a target outcome is generated in the method of regression. The generated outcome is a probabilistic interpretation examined after the intensity of the association between the source dataset is considered [2]. For instance,

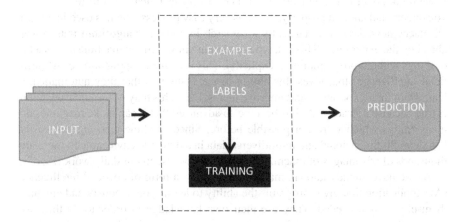

Fig. 8 Supervised learning process

regression can help predict a building's cost according to the venue, capacity, etc. The outcome has discrete values in logistic regression based on a number of independent variables.

While facing problems like non-linear and multiple decision boundaries, this approach can flounder. It is also not versatile enough for complex relationships to be recorded in datasets.

Classification

Binary Classification

It requires sorting the knowledge into grades. It is called binary classification if the supervised learning method tags input data into two separate groups. Several classifications [4] indicate the category of data into more than two different groups.

Naive Bayesian Model

An algorithm that reacts on the basis of the appearance of a certain characteristic is irrespective of the appearance of several other characteristics. The Naive Bayes algorithm's execution, considering all the sophisticated arithmetic, essentially includes keeping a list of entities with distinctive features and classes [27]. If all these statistics are collected, estimating chances and arriving at a prediction is very easy.

Decision Trees

A decision tree could be a flowchart-like design containing provisional regulation declarations, including decisions and their possible consequences. The function refers to unexpected knowledge marking. The leaf nodes refer to class labels in the tree representation, and then the internal nodes represent the features. A decision tree is sometimes used as a Boolean feature to solve issues with discrete attributes. Significant decision tree algorithms include ID3 and CART.

Random Forest Model

A prediction model is the random forest model. This involves creating a variety of decision trees and generating a hierarchy of the individual trees. Suppose you would want to guess which industry's share is going to perform well in the coming days; this algorithm will execute according to the review and previous track records of that share and current market conditions, and also will work upon the opinion of shareholders who have previously traded in these conditions; a random forest model will accomplish the objective.

Neural Networks

The purpose of this method is to combine input code, or interpret needful information by guessing and recognizing patterns. Neural networks, despite their advantages, need considerable computational resources. When there are thousands of findings, it could become difficult to suit a neural network. As analysing the logic behind their predictions is always daunting, it is often referred to as the black-box algorithm.

Support Vector Machines

A supervised learning algorithm developed in the year 1990 could be the Support Vector Machine (SVM). It draws from the hypothesis of statistical learning developed by Vap Nick. It a selective classifier because hyperplanes are removed by SVM. The efficiency is generated in the form of an appropriate hyperplane that categorizes new examples. SVMs [26] are intimately associated to and utilized in diverse sectors in the kernel framework. SVM can be used in the advancement of bioinformatics, pattern recognition, and multimedia information.

7.1.2 Pros and Cons of Supervised Learning

Many shades of supervised learning encourage peoples to accumulate and extract knowledge from previous experience. Supervised learning has demonstrated great potential in the AI domain, from optimizing recent incidents to handling actual issues. In preference to unsupervised learning, it is also a more trustworthy method, which can be computationally complex and less effective in some circumstances. Supervised learning, however, is not without shortcomings. In an IoT-based smart city architecture as implemented by authors in [21], development and progress are not possible without trust. The security of each system, sensor, and solution is not optional; it must be taken into account right from the start. For training classifiers, concrete examples are required, and decision thresholds are often overtrained in the absence of the appropriate examples. In classifying big data, one can also encounter difficulties.

7.2 Semi-Supervised Learning

The concept of semi-supervised learning comes under learning without intervention and learning with intervention. Some of the training instances are excluding training labels, yet many researchers in machine learning have found that unlabelled data can show a significant improvement in learning consistency when it is used in tandem with a minuscule percentage of labelled data. Between supervised and unsupervised learning fall semi-supervised machine learning algorithms, as they use both labelled and unlabelled data for processing, probably a decent fraction of labelled data and a big fraction of unlabelled data. The systems that use this framework are ready to improve the accuracy in learning considerably. Invariably, semi-supervised learning is chosen when skilled and specific resources are required for both the acquired labelled data to coach it/learn from it. Alternatively, it does not necessarily require substantial resources to obtain unlabelled data.

Semi-supervised learning is also very effective in the area of machine learning; its workflow is such that first the user will determine the nature of training pattern, that is, user takes decision about the training dataset which they are going to use; after the determination of training set, it needs to be related to real-world functions.

Now the next task is to find the input feature representation of the learned function; in this phase, the process will transform the input object into feature vector. Now the appropriate algorithm will be chosen for the actual execution of input data various algorithms are present like support vector machine or it may be decision trees method. Further selected algorithms should be applied on training data in order to achieve the validation set. After all these steps, finally, evaluation of the accuracy will take place and the performance of the resulting function will be measured against the test set.

These kinds of learning methods are used in different areas of application such as information extraction, spam detection and downward causation in biological system.

7.3 Unsupervised Learning

Where the information that will not be educated is neither categorized nor labelled, unsupervised machine learning algorithms are utilized. Unsupervised learning explores the way systems can infer a feature from unlabelled data to describe a secret structure. The device does not find the correct performance, but it explores the details and can draw dataset inferences to describe hidden structures from unlabelled data. Unsupervised learning algorithms take a knowledge group containing only source data and find structure within the data, such as knowledge point grouping. Therefore, the algorithms learn from raw data that has not been named, graded or categorized. Instead of responding to feedback, unsupervised learning algorithms recognize commonalities within the data and respond to each new piece of information, supporting the commonalities' existence or absence. In statistics, a core utilization of unsupervised learning is within the field of target estimation, such as discovering the function of probability density (Fig. 9).

While unsupervised learning includes several areas, data characteristics are summarized and clarified. The analysis of a cluster is the allocation of a set of observations into subsets (hierarchical clustering) in such a way that analysis within the same group are identical according to one or more predestined criteria, while analysis from different groups are different. Various clustering techniques make different assumptions about the information structure, often described by some metric of similarity and measured, for example, by internal compactness, or the similarity between members of identical groups, and the distinction between clusters. Other methods endorse the approximate density and connectivity of graphs.

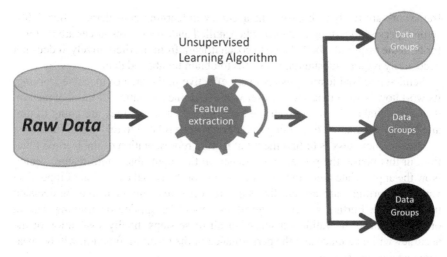

Fig. 9 Unsupervised learning process

8   Machine Learning in Malware Detection

8.1   Dataset Collection

Machine-learning classification methods need a large number of labelled executable codes for each class, but for a real-world problem such as nasty code analysis, it is very difficult to get this amount of labelled data [2]. A prolonged study process is necessary to produce these data, and some nasty executables may avoid detection within the process. Several methods are suggested to impact this problem within the full framework of machine learning. Semi-supervised learning is a type of machine-learning technique that is particularly useful for each class, if there is a limited amount of labelled data. These techniques generate a supervised classifier based on labelled data and guess the label for all unlabelled instances. Instances whose classes with a specific confidence level are expected are added to the named dataset. If such conditions are met, the procedure is repeated. The accuracy of completely unsupervised methods is enhanced by these approaches.

8.2   Features Extraction

Feature extraction [16] is a catchall concept for techniques to create correlations to fix these issues while still representing the information with adequate accuracy. Numerous machine learning experts assume that the secret to successful template matching is perfect strategy feature extraction.

8.3 Features Selection

The primary and most vital step in the current model design should be feature selection and data cleaning. You will find feature selection strategies in this chapter that you can simply use in machine learning. Feature selection [16] is the method in which you pick those features manually or automatically that most benefit your prediction variable or output during which you are curious about having irrelevant features in your data will reduce the prediction's performance and make model learn extracted attributes that are assisted.

8.4 Training of Classifier

A classification model of labelled data and appropriate assigned classification labels is acquired in one typical implementation, and an automatic classifier against training set is trained. At least one of the training samples is selected and demands confirmation/relabelling of it. A reaction identification label is obtained in response and is used to retrain the automatic classifier [23, 24].

9 Conclusion

As we found in this research, machine learning is an essential tool to combat malware. Good knowledge of machine learning and today's problems should be a distant complement.

We can avoid new and dangerous malware with the help of machine learning. Our main objective was to find a system for machine learning that typically detects the maximum number of malware samples it can, with the tough restriction of having a void false positive rate. There is a need to incorporate several deterministic exemption mechanisms in order for this system to be a part of a competitive private enterprises product. Malware detection through machine learning, in our opinion, will not replace the methods of quality detection used by anti-virus providers, but will be an additional advantage to them.

References

1. Santos, I., Nieves, J., & Bringas, P. G. (2011). Semi-supervised learning for unknown malware detection. In International Symposium on Distributed Computing and Artificial Intelligence (pp. 415–422). Springer, Berlin, Heidelberg.
2. Anderson, H., & Roth, P. (2018). EMBER: An Open Dataset for Training Static PE Malware Machine Learning Models. ArXiv, abs/1804.04637.
3. https://cuckoosandbox.org/
4. Narudin, F. A., Feizollah, A., & Anuar, N. B. (2016). A gani – Soft computing. Springer.
5. Santos, I., Devesa, J., Brezo, F., Nieves, J., & Bringas, P. G. (2013). Opem: A static-dynamic approach for machine-learning-based malware detection. In International joint confer-

ence CISIS'12-ICEUTE´ 12-SOCO´ 12 special sessions (pp. 271–280). Springer, Berlin, Heidelberg.

6. https://www.quickheal.co.in/threat-reports
7. Yang, C., Xu, J., Liang, S. et al. DeepMal: maliciousness-Preserving adversarial instruction learning against static malware detection. Cybersecur 4, 16 (2021).
8. Talukder, Sajedul. (2020). Tools and Techniques for Malware Detection and Analysis.
9. Babaagba, K. O., & Adesanya, S. O. (2019). A Study on the Effect of Feature Selection on Malware Analysis using Machine Learning. In ICEIT 2019: Proceedings of the 2019 8th International Conference on Educational and Information Technology (51–55). https://doi.org/10.1145/3318396.3318448
10. Shalaginov, A., Banin, S., Dehghantanha, A., & Franke, K. (2018). Machine learning aided static malware analysis: A survey and tutorial. In Cyber threat intelligence (pp. 7–45). Springer, Cham.
11. https://symantec-enterprise-blogs.security.com/blogs/feature-stories/symantec-security-summary-june-2020
12. Hausken, K., & Welburn, J. W. (2020). *Information systems Frontiers*. Springer.
13. Kumar, M., Punia, S., Thompson, S., Gopal, D., & Patan, R. (2020). Performance analysis of machine learning algorithms for big data classification. *International Journal of E-Health and Medical Communications (IJEHMC), 12*(4), 60–75.
14. Sharma, A., & Sahay, S. K. (2014). Evolution and detection of polymorphic and metamorphic malware: A survey. *International Journal of Computer Applications, 90*(2), 7–11.
15. Govindaraju, A. (2010). *Exhaustive statistical analysis for detection of metamorphic malware.* Master's project report, Department of Computer Science, San Jose State University.
16. Ahmadi, M., Ulyanov, D., Semenov, S., Trofimov, M., & Giacinto, G. (2016). Novel feature extraction, selection and fusion for effective malware family classification. In *ACM conference data application security privacy* (pp. 183–194). ACM.
17. Sharma, A., & Sahay, S. K. (2016). An effective approach for classification of advanced malware with high accuracy. *International Journal of Security and Its Applications, 10*(4), 249–266.
18. Bhardwaj, A., Al-Turjman, F., Kumar, M., Stephan, T., & Mostarda, L. (2020). Capturing-the-invisible (CTI): Behavior-based attacks recognition in IoT-oriented industrial control systems. *IEEE Access*, 1. https://doi.org/10.1109/ACCESS.2020.2998983
19. Shankar, A., Pandiaraja, P., Sumathi, K., Stephan, T., & Sharma, P. (2020). Privacy preserving E-voting cloud system based on ID based encryption. *Peer-to-Peer Networking and Applications.* https://doi.org/10.1007/s12083-020-00977-4
20. Stephan, T., Al-Turjman, F., Suresh Joseph, K., & Balusamy, B. (2020). Energy and spectrum aware unequal clustering with deep learning based primary user classification in cognitive radio sensor networks. *International Journal of Machine Learning and Cybernetics.* https://doi.org/10.1007/s13042-020-01154-y
21. Chithaluru, P., Al-Turjman, F., Kumar, M., & Stephan, T. (2020). I-AREOR: An energy-balanced clustering protocol for implementing green IoT in smart cities. *Sustainable Cities and Society*, 102254. https://doi.org/10.1016/j.scs.2020.102254
22. Yadav, S. P., Mahato, D. P., & Linh, N. T. D. (2020). *Distributed artificial intelligence: A modern approach* (1st ed.). CRC Press. https://doi.org/10.1201/9781003038467
23. Kumar, M., & Srivastava, S. (2018). Image authentication by assessing manipulations using illumination. *Multimedia Tools and Applications, 78*(9), 12451–11246.
24. Aggarwal, A., & Kumar, M. (2020). Image surface texture analysis and classification using deep learning. *Multimedia Tools and Applications.* https://doi.org/10.1007/s11042-020-09520-2
25. O'Kane, P., Sezer, S., McLaughlin, K., & Im, E. G. (2013). SVM training phase reduction using dataset feature filtering for malware detection. *IEEE transactions on information forensics and security, 8*(3), 500–509.
26. Shang, F., Li, Y., Deng, X., & He, D. (2018). Android malware detection method based on naive Bayes and permission correlation algorithm. *Cluster Computing, 21*(1), 955–966.

Automating Index Estimation for Efficient Options Trading Using Artificial Intelligence

Vivek Shukla, Rohit Sharma, and Raghuraj Singh

1   Introduction

In options trading, the contract's value depends on the underlying (stock or commodity) value. The present discussion is centred on options of stocks. There are two types of option trades: call option and put option. In the call option, buyer of the contract can exercise the right or drop the purchasing of stocks on the maturity date at a fixed rate. On the other side, there is put option. In the put option, the buyer of the contract can exercise the right to sell the stocks at a fixed rate or drop the contract. The other party involved is called writer or seller of contract. Contracts are usually made of different lot sizes that may vary from few hundred to thousand. For example, in a call option for some "ABC" share, the buyer may purchase the right of buying 100 shares at a rate of 30$ on 15 January 2020 by paying the cost of 700$. On maturity date, the buyer has two options: in the first option, the buyer drops the idea of purchasing because shares of underlying stock are trading at lower than 30$. Therefore, in this case, buyer loses the money (700$). In the second option, buyer may exercise the right to purchase the shares at a fixed rate because the market price of underlying stock is trading higher than 30$ say 39$. Therefore, buyer makes the profit 100 * (39 − 30)−700 = 200$.

In options trading [23], any contract may be a combination of various trades. Some may be call option, others may be put option. For example, there are multiple strategies like bull call spread, bear spread, etc. In these strategies, one might buy a call option and sell another call option of the same underlying asset. Many features

V. Shukla (✉) · R. Sharma
Department of Computer Science & Engineering, Dr. Ambedkar Institute of Technology for Handicapped U.P., Kanpur, UP, India

R. Singh
Department of Computer Science & Engineering, Harcourt Butler Technical University, Kanpur, UP, India

of option trading [24] attract researchers to develop an efficient system that draws conclusions about the movement of index or stock so that more and more accurate decisions might be made for trading. The use of AI in finance is growing rapidly, and the recent years saw many investors begin to use it in a variety of ways, though it is unclear how many of them are using it specifically for options trading. Advances in artificial intelligence (AI) have a more significant impact even in medicine and bolster the power of medical interventions and diagnostic tools as used by authors in [20]. In the financial community, Big Data is making more of a splash than a ripple. Technology is advancing at an alarming pace, with far-reaching repercussions. Industry operations are being transformed by increasing complexity and data generation, and the finance industry is no exception. AI can efficiently sift through large quantities of Big Data to generate data predictions and cost-effective energy optimization solutions to fuel Smart City technologies such as the one used by authors in [21].

2   Constraint Programming and the CLP Scheme

Complex problems can be represented concisely in constraint solving because the properties can be represented implicitly without bindings to other variables. Constraint logic programming (CLP) provides a foundation for programming language class by combining constraint-solving and logic programming concepts. CLP(R) language is implementation of constraint logic programming approach in domain of real numbers. For example, the constraint $X \geq 1, X < X + 2$ is always solvable in CLP(R) [1, 17].

A CLP(R) clause has the form as

$$p(\dots) : -p_1, \ p_2, \dots \ p_n$$

where $n \geq 0$ and p_i represents constraints or predicates.

Constraints can have many relations ($\geq, \leq, >, =$). CLP(R) has natural inference mechanism and declarative power of logic programming [18].

2.1   The Binomial Option Pricing Formula

In the model of binomial pricing, value is calculated for the underlying stock. In this analysis, it is assumed that the underlying stock's value can either move up or move down at the end of one-time unit. According to this assumption, value of underlying stock can be formulated as

$$C = \max \ (S\Delta + P, \ S - Q)$$

provided

$$uS\Delta + rP = \max\left(0,\ uS - Q\right)$$

$$dS\Delta + rP = \max\left(0,\ dS - Q\right)$$

where S represents price of stock, Δ stands for rate of change in option value with relation to change in the underlying stock, P stands for amount of currency in riskless bonds, Q stands for strike price of option contract, r is the value of interest rate $+1$, and u, d represent the up/down return rates $+1$.

The above formulas represent the constraints that can directly yield the CLP(R) rule as

Valuation (C, S, Delta, P, R, Q, Up, Down):

$$Up * S * Delta + R * P = \max\left(0,\ Up * S - Q\right),$$

$$Down * S * Delta + R * P = \max\left(0,\ Down * S - Q\right),$$

$$C = \max\left(S * Delta + P,\ S - Q\right).$$

The goal can evaluate the option price for given values of S, Q, R, Up, Down. For obtaining the value of C, CLP(R) solves several linear equations and produces answers. By iterating the evaluation rule backward for a period of time, binomial pricing model is obtained. An option value tree is computed by CLP(R) clauses using above recursion [1, 19].

2.2   The Black-Scholes Model

The formula of the Black-Scholes model is limit case of formula for binomial pricing. It is based on continuous model of option trading. It represents the theoretical price of a call option in terms of strike price (k), stock price (s), the current interest rate (risk-free) r, the time to expiration t, and volatility v for the stock calculated as square v root of variance by the following equation[2]:

Theoretical value for option price $= sN(d_1) - ke^{-rt}N(d_2)$

$$d_1 = \frac{lnln\left(\dfrac{s}{k}\right) + \left(r + \dfrac{v^2}{2}\right)t}{v}\sqrt{t}$$

$$d_2 = d_1 - v\sqrt{t}$$

where ln stands for natural log and $N(y)$ stands for cumulative normal distribution [7, 8].

3 Complexity in Options Market

Complex adaptive system theory gives a new way to study economics. This suggests that an economic system having proactive and adaptive individuals is process-based and evolving and self-organizing rather than predictable, mechanical, and established. To optimize trading, information is exchanged, and the members in the system accumulate this knowledge [15].

The market of options [25] is very complex in terms of adaptability. It depends upon the behaviour of investors and their decision-making approach. Market and investors both are affecting each other and thus changing continuously. Following is the discussion of a simulation of agent-based model.

3.1 The Architecture and Environment of Agent-Based Model

There is a mechanism very common in the market known as Continuous Double Auction (CDA). In this model, each participant is given a role that may be of buyer or of seller. In the process of trading, the seller gives a bid for a stock. This bid can be seen by all participants of the market. An investor analyses the bid given by seller. If it might maximize the gain for the investor, then bid given by the seller is accepted, and thus a transaction is completed [16, 26].

In this model, many experimental methods are chosen for a particular trading approach. Thus, this model is very helpful in simulating a continuous double auction market. This model considers the fact that investors are not aware of any information regarding valuation and cost of entity.

3.2 Agent Strategies (Zero Intelligence Plus Model)

ZIP is machine learning-based approach that determines optimal strategy by experience. It adapts according to open auction market environment which helps to make decisions about strategies to be followed. Any agent i in this model has profit margin $\mu_i(t)$ and limit profit α_i. An agent can submit a bidding price $p_i(t)$ as [3]

$$p_i(t) = \alpha_i \left(1 - \mu_i(t)\right)$$

and the seller can submit asking price as

$$p_i(t) = \alpha_i \left(1 + \mu_i(t)\right)$$

An agent i is selected randomly and asking or bidding price $p_i(t)$ is the same as equal to the current price in market $q_i(t)$. If the asking or bidding price is lesser than (or more than) the current price in the market, then the transaction may happen. Whether a transaction happens or not, all agents update the profit margin $\mu_i(t)$ according to learning algorithm [9, 10].

3.2.1   Model Parameters

Every agent is allotted some parameter in this model – parameter for learning rate β_i, profit margin $\mu_i(t)$ and dynamic parameter ψ_i. Genetic algorithm is used for the optimization of transaction process. Many issues are considered whether the underlying price goes up or down, movement of option price, asking or bidding price is less or more than the current market price, etc. According to these issues, market state is analysed, and on the basis of market state, strategies are formed to either increase or maintain the profit.

If the profit increases for a particular strategy, then it is believed by agent that strategy was optimal that market state and thus the probability of using that strategy again increases. On the other hand, if the profit decreases, it leads the agent to believe that strategy was not appropriate in that market state and the probability of using that strategy again is reduced. Thus, using this learning approach, the agent gets to choose strategies for making profit.

3.2.2   Optimizing Strategies Using Monte Carlo Simulation

In the market, strategies get affected by current market price and the future prices of stock. In Monte Carlo simulation, random sampling is used for state variables. The first random sample path is generated for stock price under consideration as $p1, p2, p3, \ldots pT$ over a relevant time horizon. Assuming the Brownian motion, the price for the underlying stock is [3]:

$$\frac{dS}{S} = \mu dt + \beta dz$$

where dS represents increment in stock price, μ stands for asset return, β represents the price volatility, dz stands for Wiener increment, where z is a Wiener process. Assuming the risk of neutral condition, the above formula can be rewritten as:

$$\frac{dS}{S} = \left(\mu - \frac{\beta^2}{2}\right)dt + \beta dz$$

In the above derivation of formula, assume the time interval is divided into X sub-intervals having length $\Delta t = T/X$. Thus, a new formula can be written as

$$\ln S_i - \ln S_{i-1} = \left(\mu - \frac{\beta^2}{2}\right)\Delta t + \beta\sqrt{\Delta t} \cdot \varepsilon_i$$

where ε_i represents standard normal distribution. If underlying price S_0 is given, then price S_i can be calculated as

$$S_i = \exp(\ln S_0 + \left(i \cdot \left[\left(\mu - \frac{\beta^2}{2}\right)\Delta t + \beta\sqrt{\Delta t} \cdot \varepsilon_i\right]\right)$$

Simulating the above formula many times and averaging it, underlying price can be calculated [13, 14].

3.2.3 Process of Strategy Formation

As it is clear from the above analysis, strong relationship exists between underlying and option price. Given the price distribution of underlying, option's life is divided into several small units. Random sampling is done on samples of distributions. With the help in simulation of price changes, movement of price's path can be predicted. Thus, value for the price of option can be calculated. By taking a random sample from the set of terminal values, the result can be analysed.

Model of equilibrium for rational expectations can be very helpful here. Over a particular period, it is assumed that $x_{i,j}$ stands for shares of underlying stock that are holed by buying agent I who is willing to accept the writer's asking price and $y_{i,j}$ stands for the writer who is willing to accept the bidding price. $\psi_{i,j}$ represents the demand for underlying stock, then the following relationship holds among them,

$$x_{i,j} = \psi_{i,j} - \psi_{i,j-1} \qquad \text{if } \psi_{i,j} \geq \psi_{i,j-1}$$

Otherwise, $x_{i,j}$ is 0

$$y_{i,j} = \psi_{i,j-1} - \psi_{i,j} \qquad \text{if } \psi_{i,j} < \psi_{i,j-1}$$

Otherwise, $y_{i,j}$ is 0

$$X_t = \sum_N^{i=1} x_{i,j}$$

$$Y_t = \sum_N^{i=1} y_{i,j}$$

where X_t denotes number of holders and Y_t stands for number of writers. The α represents the difference function between X_t and Y_t. α can be expressed as regulation speed of price as

$$\alpha(Xt - Yt)) = \{\tan h(\alpha(Xt - Yt)), Xt \geq Yt \tan h \tan h(\alpha(Xt - Yt)), Xt < Yt$$

and

$$\tan h(x) = \frac{e^x + e^{-x}}{e^x - e^{-x}}$$

The price prediction for underlying price can be formulated as [3]

$$S_{t+1} = S_t \left(1 + \alpha \left(X_t - Y_t\right)\right)$$

From other paths, different random samples are taken. After repetition of this procedure hundreds of times, the set of options at a particular time is obtained. Averaging over different sample paths, calculation of option price is done. Agents tend to make a decision by comparison of market and calculated option price. This helps also in assessing the profit scenario for trader. Thus, the agents can determine whether to write or hold the options. Repetition of this process very much reflects the market situation.

To choose the opening prices, a method can be used which optimizes the potential profit of the makers of the market. This method can be used to offset the imbalances in public order with few limits on the allowed deviation for implied volatilities. In this method, set of implied volatilities σ over some range of possible values of implied volatilities.

Corresponding prices are examined for each element in this set. This helps in determining the side responsible for imbalances in the public order. For each imbalance in the buying side, an upward move is followed from price p_i in the set σ_i. For each price greater than p_i, product of buying imbalances is calculated, which helps determine the profit potential in the market.

A similar process is followed for the imbalances in selling side. There is a difference when followed for selling side; instead of moving forward, stepping down is followed in prices. A price can be found in each series for which profit potential can be maximum but a bound which is exchange specified for a deviation which is maximum from price p_i should also be considered.

Opening prices are picked by exploring over the sets for implied volatilities σ. The above procedure results in sets of prices that helps in maximizing the traded volume.

3.2.4 Enhancement of Model Based on Delta-Gamma Parameters

Option price can be calculated using Black-Scholes formula as

$$C_t \triangleq C^{(BS)}\left(S_t, T-t, \sigma_t, r_t, K\right)$$
$$= S_t \phi(d_+) - K \exp\left(-r_t - (T-t)\right)\phi(d_-)$$

where S_t is the price of underlying, T represents maturity time, implied volatility is represented by σ_t, r_t whose value is greater than zero represents the prevailing rate of interest, K represents the striking price of option, ϕ represents the cumulative distribution function and d_+, d_- are described as

$$d_+ / d_- = \frac{1}{\sigma t}\left(\ln\left(S_t / K\right) + r_t\left(T-t\right)\right)$$

Option price can be represented as

$$V_t \triangleq (S_t - S_0)/S_0$$

where V_t stands for linear return of stock price during the time $(0,t)$ and S_0 stands for underlying price at beginning.

According to risk or position (in both cases) management, change in option price can be represented as

$$C_t - C_0 = C^{(BS)}\left(S_t,\ T-t,\ \sigma_t,\ r_t,\ K\right) - C^{(BS)}\left(S_0,\ T,\ \sigma_0,\ r_0,\ K\right)$$

Some modifications can be made as

$$C_t - C_0 = C^{(BS)}\left(S_t,\ T-t,\ \sigma_t,\ r_0,\ K\right) - C^{(BS)}\left(S_0,\ T,\ \sigma_0,\ r_0,\ K\right)$$

assuming that interest rate is uniform mostly.

On assuming that volatility is also constant, then above equation can be modified further as

$$C_t - C_0 = C^{(BS)}\left(S_t,\ T-t,\ \sigma_0,\ r_0,\ K\right) - C^{(BS)}\left(S_0,\ T,\ \sigma_0,\ r_0,\ K\right).$$

It can be observed that option price calculation based on approximation of delta gamma parameters drops as change in underlying stock price increases for all the parameters which are considered here. This behaviour becomes more important as the value of volatility decreases (Fig. 1).

The reason for the above changes is that approximation is made at beginning while approximation becomes more accurate if it is made at t time. This modification leads to a new concept called extended delta gamma approximation.

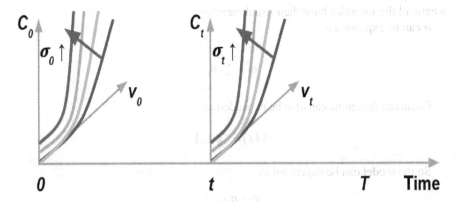

Fig. 1 Change in implied volatility and asset price [2]

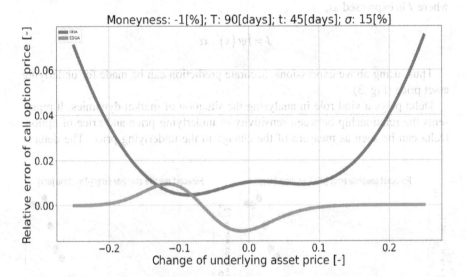

Fig. 2 Comparison of EDGA and DGA [2]

Thus, according to extended delta gamma approximation, the above equations get modified as

$$C_t - C_0 = C^{(BS)}\left(S_t^*, T-t, \sigma_t, r_0, K\right) - C^{(BS)}\left(S_t^*, T, \sigma_0, r_0, K\right)$$

Using functional data analysis, change in stock prices can be predicted (Fig. 2).

The performance relies on accuracy of prediction in the changes of underlying. In this method, modelling the scalar response as linear model as

$$X = \alpha + \int_T^0 X \cdot \alpha \cdot ds + \varepsilon$$

where all the variables have their usual meaning.

α can be expanded as

$$\alpha = \sum_{K}^{(k=1)} b_k$$

Covariate functions can also be expanded as

$$X(s) = C\psi(s)$$

So the model can be expressed as

$$X = \beta + CJ$$

where J is expressed as,

$$J = \int \psi(s) \,.\, \alpha$$

Thus, using above expressions, accurate prediction can be made for underlying asset price (Fig. 3).

Delta plays a vital role in analysing the situation of market dynamics. It represents the relationship between sensitivity of underlying price and price of option. Delta can be seen as measure of the change in the underlying price. The delta is

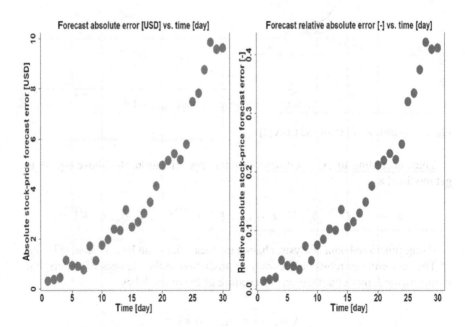

Fig. 3 Absolute vs relative error forecast [2]

calculated by change in price of option divided by change in the underlying price [11].

Concept of gamma is used to study the delta values. Gamma is used to measure the closeness of market price to the option price. As the closeness of these two prices increase, value of gamma goes higher. Gamma is also related to expiration date. As the maturity date comes closer, value of gamma increases as well.

Change in call option price can be predicted using delta and gamma approximation as

$$C_t - C_0 \approx \Delta v_t S_0 + \Gamma \left(v_t S_0 \right)^2$$

where C_t and C_0 are option prices and S_0 is underlying price, v_t is linear change in underlying asset price and Δ, Γ are parameters related to probability function and distributive function [12].

4   Estimating the Price Movement with Normal Distribution Curve of Underlying Asset

As the above calculation suggests, the price of underlying can be assessed by random sampling taken over different samples. Normal distribution curves can help to conclude the price movement over a specific time period. The Gaussian distribution curve is based on the central limit theorem. According to this theorem, random variables are identically distributed and independent when calculated average values show the same Gaussian (normal) distribution [4] (Fig. 4).

As it can be seen from the above diagram that within the first standard deviation, roughly 70% variation in underlying price occurs. If the transactions are based on a

Fig. 4 Graph showing the behaviour of an arbitrary stock [6]

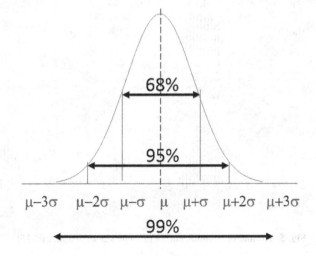

strike price which are very close to the values in first deviation, then probability of market price ending on the price higher or closer to option value increases.

The learning algorithm can be modified in such a way to exhibit these relations of option price of assets. Figure 2 shows the behaviour of some most commonly traded stocks and indexes. It is evident from figure that variation in prices of stocks are showing the similarity between Gaussian distribution and their price variation over a period of time. Strike price vary according to moneyness of options. If the prices are chosen according to precalculated values based on standard deviation and variance, then it increases the probability of profit in the transaction (Fig. 5).

The Gaussian distribution curve for asset price can be used to predict price changes of the underlying. The kurtosis and skewness helps us to determine how different is the behaviour of the underlying asset from Gaussian distribution. Zero skewness curves are very similar to ideal distribution, and thus prediction is more accurate. Negative or less than zero skewness indicate that distribution is not even and biased toward the left area. Similarly positive or more than zero skewness value

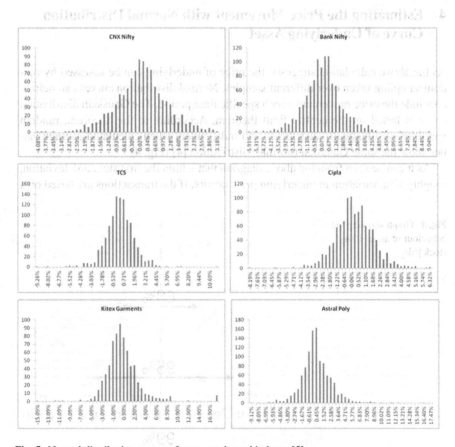

Fig. 5 Normal distribution curves of some stocks and indexes [5]

represents flatness of distribution in the right side. The more symmetric the curve, more accuracy can be achieved in predicting results.

As it can be concluded from above examples of normal distribution curves, if price estimation is between the first standard deviation, it can be used to optimize the profit because price movement is supposed to be under this range unless the market shows very strange behaviour. This calculation can be more accurate using the Monte Carlo simulation based on option pricing formula.

The planet is currently producing 2.5 quintillion bytes of data per day, which represents a once-in-a-lifetime resource for encoding, analysing, and using the data in useful ways. Machine learning and algorithms are rapidly being used in securities trading to process large amounts of data and make forecasts and judgments that humans cannot. Accurate insights into corporate decision-making models are critical in finance and trading. Humans have traditionally crunched numbers and taken choices based on inferences derived from measured risks and patterns. Computers have taken over this role today. They can compute on a large scale and pull data from a variety of sources to arrive at more precise results almost instantly. As implemented by authors in [22], even AI technologies provide promising solutions for end-to-end optimization of wireless networks. Financial analytics is no longer limited to the analysis of markets and price behaviour; it also encompasses the values that influence prices, as well as social and political patterns and the determination of degrees of support and opposition.

References

1. Lassez, C., & Mcaloon, K. (1987). Constraint logic programming and option trading. *IEEE Intelligent Systems, 2*, 42–50.
2. Jun, J.-Y., & Rakotondratsimba, Y. (2020). *Hedging option contracts with locally weighted regression, functional data analysis, and Markov chain Monte Carlo techniques.* IEEE.
3. Jie, Z., Jing, C., Dongsheng, Z., & Quan, Z. (2009). *An agent-based simulation model of options market.* IEEE.
4. https://www.investopedia.com/terms/n/normaldistribution.asp
5. https://zerodha.com/varsity/chapter/volatility-normal-distribution
6. https://sphweb.bumc.bu.edu/otlt/MPH-Modules/PH717-QuantCore/PH717-Module6-RandomError/PH717-Module6-RandomError5.html
7. MacBeth, J. D., & Merville, L. J. (1979). An empirical examination of the Black-Scholes call option pricing model. *The Journal of Finance, 34*(5), 1173–1186.
8. Raju, S. (2012). Delta gamma hedging and the Black-Scholes partial differential equation (PDE). *Journal of Economics and Finance Education, 11*(2), 51–62.
9. Meucci, A. (2008). Fully flexible views: Theory and practice. Fully flexible views: Theory and practice. *Risk, 21*(10), 97–102.
10. Cleveland, W. S., & Devlin, S. J. (1988). Locally weighted regression: An approach to regression analysis by local fitting. *Journal of the American Statistical Association, 83*(403), 596–610.
11. Castellacci, G., & Siclari, M. (2003). The practice of delta-gamma VaR: Implementing the quadratic portfolio. *European Journal of Operational Research, 150*(3), 529–545.

12. Ramsay, J. O., & Silverman, B. W. (2007). *Applied functional data analysis: Methods and case studies*. Springer.
13. Carlo, C. M. (2004). Markov chain monte carlo and gibbs sampling. *Lecture Notes for EEB, 581*.
14. Gilks, W. R., Richardson, S., & Spiegelhalter, D. (1995). *Markov chain Monte Carlo in practice*. Chapman and Hall/CRC.
15. Yadav, S. P. (2020). Vision-based detection, tracking and classification of vehicles. *IEIE Transactions on Smart Processing and Computing, SCOPUS.*, ISSN: 2287-5255, *9*(6), 427–434. https://doi.org/10.5573/IEIESPC.2020.9.6.427
16. McMillan, L. G. (1986). *"Options as a strategic investment," in New York Institute of Finance* (2nd ed.). Prentice-Hall.
17. Yadav, S. P., Agrawal, K. K., Bhati, B. S., et al. (2020). Blockchain-based cryptocurrency regulation: An overview. *Computational Economics*. https://doi.org/10.1007/s10614-020-10050-0
18. Heintze, N. C., et al. (1986). *The CLP(R) programmer's manual*. Tech. Report 73. Monash University.
19. Steele, G. L., Jr., & Sussman, G. J. (1979). *Constraints* (pp. 208–225). ACMSIGPLAN STAPL APL Quote Quad.
20. Stephan, P., Al-Turjman, F., & Stephan, T. (2020a). Severity level classification and detection of breast cancer using computer-aided mammography techniques. *Wireless Medical Sensor Networks for IoT-Based EHealth*, 221–234. https://doi.org/10.1049/pbhe026e_ch13
21. Chithaluru, P., Al-Turjman, F., Kumar, M., & Stephan, T. (2020). I-AREOR: An energy-balanced clustering protocol for implementing green IoT in smart cities. *Sustainable Cities and Society*, 102254. https://doi.org/10.1016/j.scs.2020.102254
22. Stephan, T., Al-Turjman, F., Joseph, K. S., Balusamy, B., & Srivastava, S. (2020). Artificial intelligence inspired energy and spectrum aware cluster based routing protocol for cognitive radio sensor networks. *Journal of Parallel and Distributed Computing*. https://doi.org/10.1016/j.jpdc.2020.04.007
23. Punia, S. K., Kumar, M., & Sharma, A. (2021). Intelligent data analysis with classical machine learning. In S. S. Dash, S. Das, & B. K. Panigrahi (Eds.), *Intelligent computing and applications* (Advances in intelligent systems and computing) (Vol. 1172). Springer. https://doi.org/10.1007/978-981-15-5566-4_71
24. Wu, J. M. T., Wu, M. E., Hung, P. J., et al. (2020). Convert index trading to option strategies via LSTM architecture. *Neural Computing and Applications*. https://doi.org/10.1007/s00521-020-05377-6
25. Chen, A. P., Chen, Y. C., & Tseng, W. C. (2005). Applying extending classifier system to develop an option-operation suggestion model of intraday trading – An example of Taiwan index option. In R. Khosla, R. J. Howlett, & L. C. Jain (Eds.), *Knowledge-based intelligent information and engineering systems. KES 2005* (Lecture notes in computer science) (Vol. 3681). Springer. https://doi.org/10.1007/11552413
26. Aggarwal, A., Alshehrii, M., Kumar, M., Alfarraj, O., Sharma, P., & Pardasani, K. R. (2020). Landslide data analysis using various time-series forecasting models. *Computers and Electrical Engineering, 88*(2020), 106858. https://doi.org/10.1016/j.compeleceng.2020.106858

Artificial Intelligence, Big Data Analytics and Big Data Processing for IoT-Based Sensing Data

Aboobucker Ilmudeen

1 Introduction

The recent technologies such as artificial intelligence, cloud computing and Internet of things have empowered the analytics more dominant. Scholars claimed that IoT, big data analytics, AI and cloud computing are cutting-edge developments that have transformed the globe [5]. These techniques are mutually supportive and produce a number of cross-disciplinary and interdisciplinary fields of study and application. These comprise not only ICT applications, but also all types of systems in our community, covering healthcare, business industry, production, entertainment, education sector, and the environment [5]. Similarly, the big data analytics, artificial intelligence and machine learning continue to see use in various management fields [8].

The volume of data in today's business environment is astounding. Big data, though, provides a broad variety of options for enterprises, whether used separately or alongside current conventional data. Statisticians, analysts, academics and entrepreneurs, employees will use these emerging databases for sophisticated analytics that offer greater perspectives and fuel creative big data solutions. Some of the important approaches include data mining, text classification, prediction, visual analytics, artificial intelligence, statistics and language processing. The unlimited storage capability and cloud infrastructure related to cloud computing contribute to a modern domain in big data and big data handling [5]. Today, the business firms around the world are seeking at big data-enabled computational approaches including the IoT, smart assistants, artificial intelligence, machine learning, deep learning, intelligent robots, content analytics, neuroscience business models, etc. [25].

A. Ilmudeen (✉)
Department of Management and Information Technology, Faculty of Management and Commerce, South Eastern University of Sri Lanka, Oluvil, Sri Lanka
e-mail: ilmudeena@seu.ac.lk

© The Author(s), under exclusive license to Springer Nature Switzerland AG 2022   247
F. Al-Turjman et al. (eds.), *Transforming Management with AI, Big-Data, and IoT*, https://doi.org/10.1007/978-3-030-86749-2_14

The data analytics and big data are becoming ever more significant with the connection-ready devices and the IoT-related technologies. Recent technological developments allow for increased computational capacity for smart devices that produce real-time data rapidly [13]. The IoT gathers large volume of data that includes information about millions of things. To be able to analyse this data, it requires the aid of artificial intelligence. This form of technology helps them to see meaning and to know the contextual relations and trends that explores hidden business insights through analytics. IoT is enabling big data analytics to make instantaneous decisions. If we look deeply into it, there is a clear connection between big data and IoT.

The IoT has performed a large part in today's business scenario. For example, IoT lets e-commerce companies maintain up-to-date inventory data that minimize overstocking in warehouses by using IoT-centric sensors. In addition, IoT sensors help the monitoring of orders from original placement to final transport that guarantee that they are not scattered and delivered on time with appropriate worth. IoT encompasses technology such as RFID, cloud computing, cameras, GIS/GPS, visualization, virtual reality, and augmented reality [18]. These technologies are related to enable the business intelligence and decision-making support system. With the aid of IoT, the Business Intelligence and Decision Support System focuses on tailored consumer promotion in such a way as to find the purchasing trend through web searching on the market for their expected customers. IoT devices such as smartphones, laptops, wearable sensors or industrial sensors may be used for banking operations in the financial sector. Market intelligence can be used by many banking clients to obtain secret insights into deposits, savings, promotional events and transactions.

1.1 Big Data

The rapid development of big data focuses on new ways of capturing, storing, exchanging, saving, analysing and displaying knowledge. The word Big Data is described differently in the literature [14]. Big data is typically processed by artificial intelligence, and both of these are used ever more in healthcare research (e.g., [12, 22].). By its very nature, Big Data is an incredibly huge, complex and continuously changing domain that has the ability to transform, but can also influence our daily routine. Big data is broadly meant to be handled by artificial intelligence technique and its sub-discipline, machine learning. However, it may also be possible to handle "small" data sets by using artificial intelligence and machine learning.

1.2 Internet of Things (IoT)

With recent rapid developments in computer and networking technology, a new stage of growth in the digital era "Internet of Things" came into being. The IoT signifies the ever-growing physical objects that enable certain artifacts to communicate independently and smartly, including digital sensors, cars, buildings and other objects integrated with sensors, applications, detectors, actuators and network communication. IoT is the cornerstone of smarter network services including smart house, smart environment, energy, traffic control and healthcare. In order to strengthen economies, community and lifestyle, the IoT-based smart services offers enormous markets and opportunities. The applications of IoT devices have an effect on many different facets of citizen's life. The Internet of Things integrates sensors, applications, smart watches, wearables, smartphones, thermostats, voice-enabled devices, traffic lights, train vans, automobiles, and many more.

1.3 Artificial Intelligence

Artificial intelligence is the capacity of machines to conduct activities that typically require human intelligence [2]. In recent times, businesses have recognized the importance of integrating new computational methods, relying on large data and seeking to incorporate deep learning and artificial intelligence techniques into better data processing. Artificial intelligence-driven big data analytics capture more useful knowledge from which companies can develop their decision-making skills [8]. The growing amount and complexities of business data have led to the commercial deployment of artificial intelligence. The scholars concluded that the use of artificial intelligence and data analytics will include a variety of potentials to promote service offerings, efficiency and results [9].

Artificial intelligence researchers focus on deep learning and natural language processing to help computers understand correlations and assumptions. Artificial intelligence finds its presence all over the world in today's new knowledge age. Artificial intelligence systems have emphasized to boost the expertise of market analytics and intelligence. Artificial intelligence in business is increasing that are able to refine computer algorithms to recognize patterns and observations into large volumes of data and to make fast strategic decisions that are expected to remain successful in real time. Artificial intelligence and business intelligence provide superior and efficient outcomes of strategic decision [1]. Similarly, the significance of big data analytics, artificial intelligence, and machine learning has been at the front of research in diversified field such as in operations and supply chain management [8]. IT applications that incorporate artificial intelligence and modern cloud infrastructure technology would allow logistics professionals to develop their existing order management more efficiently [17]. In healthcare sector, for instance, artificial intelligence can display clinical pictures easily and accurately to take timely decisions.

2 Applications of Artificial Intelligence

Most businesses probably use some form of artificial intelligence, whether or not they know it. Accordingly, the machine learning, deep learning, automation, robotic process, and other types of artificial intelligence are integrated, allowing users to refine and simplify their business processes. Artificial intelligence technique has been applied in various field including healthcare [7], logistics industry (e.g., [17].). Artificial intelligence and high-tech cloud computing technology can make order handling more cost-effective in the logistic and supply chain [17]. Further, the convergence of technologies such as cloud computing, distributed artificial intelligence, middleware, network servers and database technology supports these kinds of architecture criteria [19].

Applications such as smart transportation networks, strategic partnership, logistics and decision support in industry have arisen from big data, IoT, artificial intelligence with IoT [18]. Techniques like artificial intelligence and big data have been applied in the field of healthcare, especially, for cancer treatment [7]. Moreover, the artificial intelligence and machine learning in the past decade has had an immense influence in the fields like manufacturing, telecommunication, development and healthcare [7].

3 Applications of IoT

Today, the diligences of IoT in the intelligence warehouse are to handle orders, grouping of items, distribution schedules, and record keeping using RFID, sensor, and wirelessly enabled objects successfully. The IoT sensor is used to track the order from the beginning to the end, ensuring that the requested goods is not broken and arrives on time. This would reduce the cost and time associated with delivering orders to clients. Intelligent logistics in the e-commerce and manufacturing industries, powered by IoT and Industry 4.0, allows for flexible manufacturing, lean production, and e-commerce expansion [18].

Material handling, automation, production scheduling, quality control, defect detection, maintenance, and machinery supervision have all seen significant advances as a result of Industry 4.0. In many aspects, IoT devices play a critical part in Industry 4.0's smart manufacturing process. IoT targets clients for personalised marketing in a method that recognises the buying trend through Internet browsing in order to sell to the desired customers.

In this line of thought, IoT serves as a strategic partner, introducing e-commerce to the next generation. In the financial industry, IoT devices such as smartphones, laptops, wearable sensors and industrial sensors might be used for banking operations.

Banking executives, for example, may monitor consumers using smartphones, tablets, and other digital devices, as well as bank retail sites [26].

4 Technologies and Methods Supporting Big Data and IoT and Artificial Intelligence

4.1 Deep Learning (DL)

Deep learning is a modern aspect of artificial intelligence, and has recently shown the possibility of increasing the performance of IoT big data analytics [2]. The latest increase in Internet of Things data has contributed to the emergence of creation of real-time data collection, along with the computer learning, deep learning, and computing technologies. The DL algorithms enable scholars to process huge quantities of fresh data instantaneously with superior precision and greater performance. In several fields, deep learning has functional and critical implications, including robotics, language processing, detection of images and expression, drug development, enhanced clinical diagnostics and precision medicine [7].

4.2 Cloud Computing

In cloud computing, monitoring, management and review of data takes place. Using Cloud infrastructure allows for more scalable, low-cost, consistent, durable, and reliable functionalities [23].The cloud computing provides high reliability, scalability and autonomy required for next-generation Internet of Things applications [11]. Cloud computing has benefited from IoT paradigm to dynamically deliver many new services, and IoT-based cloud computing will be expanded to create new services and applications in the digital world [15]. The cloud-based centralized storage is used to handle large volume of big data more effectively for real-time data processing, data analytics, diagnosis, prediction and visualization [27]. However, cloud computing has the limitations such as high latency, mobility, overhead communication and awareness of the location while handling the unprecedented amount of big data in real time [30]. The technique called fog computing is implemented at the edge of the network to overcome the aforementioned limitations in order to offer better services to the end users.

4.3 Machine Learning

Machine learning is the best way to take advantage of the hidden observations and trends from the vast volume of data set with the least human guidance support [6]. Machine learning involves a number of methods, such as predictive analytics, data mining, identification of patterns and different modelling. Through executing predictive and prescriptive analytics to support intelligent clinical facilities, the healthcare industry is keen to use the implementations of machine learning approaches in actionable knowledge bases.

Machine learning is a type of artificial intelligence that allows machines to learn without being explicitly taught and to utilise past outcomes to enhance future results, according to researchers [21]. Machine learning approaches are used in healthcare for the prediction of disease incidence and reasoning, decision support for medical surgery or treatment, the extraction of healthcare information, the study of different health data and the discovery of medicines [24]. For example, various machine learning techniques are used for mining from large-volume data sets; decision trees, support vector machines, neural networks, reduction of dimensionality, etc. [6].

4.4 Computational Techniques

Traditional health data processing techniques have been ineffective due to their inability to manage large amounts of complex data [10]. Huge data is utilized solely for analytics, which involves mining big data for information and key insights. Media, cloud, online, IoT sensors and databases are just a few of the data sources that may be utilised to collect massive amounts of big data [16]. The use of intelligent agents in the healthcare sector includes the recovery of health information from big data, disease diagnostic decision support systems, organising and arranging activities for physicians, nurses, and patients, exchange of medical information, medical image processing, automation, simulations, bioinformatics, medical data management, and health decision support systems [24]. Different data mining processes can be used in healthcare to extract efficient information, including classification, association rule mining, regression, clustering, detection, analysis, decision trees, and visualisation [16, 20]. Data mining is the practice of extracting patters and connections that can produce information or observations from databases or large data sets [16].

5 Big Data Analytics

Big data analytics is a type of modern analytics that includes advanced capabilities including predictive models, mathematical algorithms and what-if analysis driven by analytics systems. The introduction of cloud computing and big data analytics has made it easier to process huge volumes of data and business transactions [29]. Now the big data analytics and cognitive computing have been focused for improved decision-making [25]. Big data analytics techniques enable data scientists, data analysts, statistical modelers, mathematicians and other analysts to explore increasing quantities of organized transaction data, and other types of data that are sometimes left unexploited by traditional BI and analytical systems. This involves a mixture of semi-structured and unstructured data. For instance, Internet clickstream data, web server logs, social networking site information, consumer email text and survey answers, cell phone records, and computer data are collected by IoT.

By integrating the big data analytics with an artificial intelligence algorithm, it facilitates to analyse and observe the likelihood thresholds for a wide range of patients and DVH [22]. The combination of big data analytics with machine learning can help managers with decision-making and to predict unanticipated losses brought about by defective manufacturing processes [28] (Fig. 1).

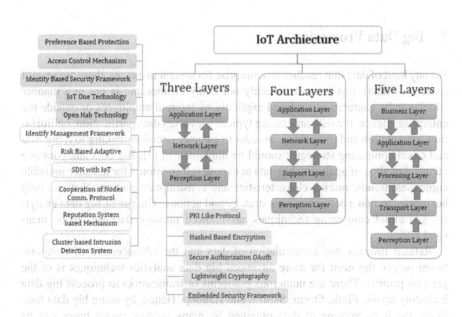

Fig. 1 Architecture and security mechanisms for IoT applications

6 IoT-Based Sensing Data

The IoT creativity has made it easy to link everything to the Internet. Now, nearly all artifacts, for instance, homes, business firms, warehouses, and even towns, are linked to the systems of network to gather data and exploit it for numerous purposes. In order to change daily life, the IoT connects everyone with new forms of resources. With this the modern technologies like cloud computing and big data are also joined. IoT sensor data are used to manage data in the smart building environment [23]. Accordingly, the sensors are typically tools that sense and react to environmental inputs that arise from different sources, for instance, bright, humidity, movement and pressure. These sensors produce useful data and these data can be exchanged with other linked devices and systems. The sensors are really important as they perform crucial operations for business enterprises. For instance, sensors can notify the possible issues before they create big losses or problems. Hence, the businesses can get ready the pre-arrangement for maintenance and prevent costly interruptions. The sensor data is able to support business decision makers to understand important patterns and produce well-versed decisions based on the situation.

Though the IoT has been around for decades, it is now sophisticated enough to go popular through sectors that are revolutionizing itself with modernization in digital era. Now the IoT-based sensors let enterprises to build innovative technological solutions and change their business models. There is a great importance in IoT automation from companies who want to do their jobs with a minimum personnel disruption and a decreased loss of efficiency.

7 Big Data Processing

For any kind of data that reaches an enterprise (in several instances, there are several sources of data), this is most certainly not really cleaner and is not in a manner which can be captured or analysed explicitly by internal employees or outside the enterprise. Hence, the data processing typically involves data purification, optimization, aggregation and processing. In order to enhance decision-making IoT, big data and cloud computing strategies should be implemented [3]. Big data analytics is a nuanced method of analysing big data to expose information – for instance invisible anomalies, trends, market characteristics and consumption patterns that can help businesses to make better business strategy and actions. The use of big data analytics and cloud computing techniques has allowed massive data sets to be managed [29].

Related towards this accumulation of data and the advancement of computational power, the need for more advanced big data analytics techniques is of the greatest priority. There are numerous platforms or frameworks to process big data including Spark, Flink, Storm, Samza, and Hadoop. Hence, by using big data handling, the large volume of data obtained by many heterogeneous bases can be managed and viewed by effective means, thereby allowing managers to make superior decisions [28].

8 Architectural Design of Big Data Analytics and IoT-Based Sensing Data and Processing

8.1 Data Collection Sources

The big data comprises of large volume, and heterogeneous sources of data including structured, unstructured and semi-structured data. The terms structure, unstructured, and semi-structured are frequently stated in the context of data and analytics. Simply, we can define the structured data as the data within a database or some kind of data management application. This structured data is systematized into an organized warehouse that is usually a database. The unstructured data can be defined as the data that is not pre-arranged or does not have predefined data model; hence, it is not a good match for a conventional relational database. The semi-structured data refers to the data that does not exist in a relational database, but it has some organizational features to analyse. This data can be stored in the relationship database using some method, for instance, XML data.

8.2 Big Data Analytics and Processing

The big data analytics handle large volumes of data to discover unseen trends, market tendencies, buyer preferences, associations and other intuitions. Hence, the enterprises can conduct quantitative and qualitative analysis with less time, money and personnel resources by using big data processing techniques.

8.3 Big Data Analytics Platforms and Tools

Big data analytics tool offers intuitions from huge data sets obtained from big data collections. These platforms and tools support enterprises tendencies, patterns, and association in data and synthesize the information into visual images, reports, and consoles of understandable form. In the industry, various big data analytics tools are available such as Hadoop, Talend, MongoDB, Spark, Kafka, Storm, Cassandra.

8.3.1 Big Data Analytics Techniques

This includes various big data analytics techniques such as artificial intelligence, deep learning, machine learning, data mining, text mining, neural networks, and intelligence agent. The big data that resides in the big data analytics platforms and tools can be analysed by using these techniques. These sophisticated techniques and efficient algorithms can interpret the collected big data. As a results, these techniques convert the data into more meaningful files (Fig. 2).

8.3.2 Big Data Processing Framework

The big data framework was created as many enterprises failed to incorporate a good big data practices into their organization, while the advantages and business cases of big data are obvious. The big data processing framework bring the advantages including big data structure and capabilities, big data-driven firm; it is vendor-free, hence, it can be used irrespective of technology, tools, or common platform as it can be applied through functional areas or country boundaries, and it can detect core and measurable capabilities for enterprises. In this proposed architecture the big data can be processed and handled to generate meaningful output files such as reports, OLAP, queries and mined data.

8.4 IoT Sensing Data

Today, the amalgamation of sensor-based communication technologies with the big data analytics made possible to apply IoT-based applications to enhance the quality of life. The motive behind the increasing usage of IoT devices is that they can make human life more comfortable [4]. Hence, the IoT sensors and devices can collect the data from various sources such as home, smart city, industry, etc.

9 Limitations and Challenges in IoT-Based Sensing Data

In view of advances in the field of medical image diagnosing, there are complexities like data and object complexity and validation issues [7]. Prior studies highlighted the challenges associated with IoT, and these can be addressed in the future research direction, for instance, poor management, identity management, trust management and policy, security, storage, authentication, authorization, securing network, privacy, data communication [4].

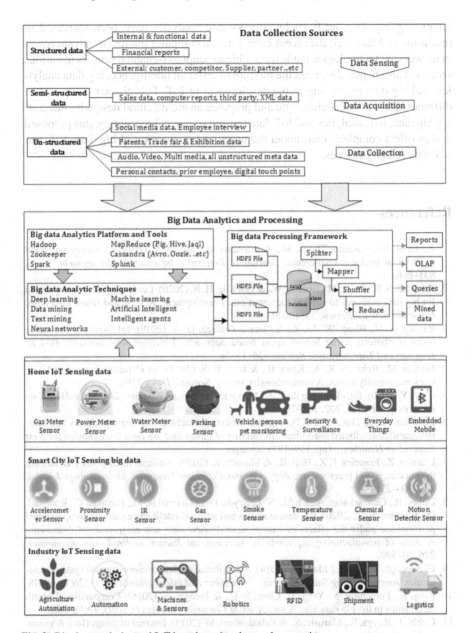

Fig. 2 Big data analytics and IoT-based sensing data and processing

10 Conclusion

The recent years have seen a rapid growth in the size, range, speed, and worth of large volume of data. Hence, the augmented applications of IoT devices by integrating big data, artificial intelligence and analytics have been increased in recent times. Moreover, the recent emergence in computer platforms, advancement in networking technologies made it possible to adopt IoT-enabled sensors in various sectors. This chapter discusses the mixture of artificial intelligence, big data analytics, and big data processing in the development of IoT. This chapter comprises of different layers in IoT architecture and proposes an architectural design by integrating big data, IoT, analytics and IoT data sensing sources. Accordingly, this proposed design offers complete ideas about the usage and application of big data, IoT, and IoT data sensing sources.

References

1. Aruldoss, M., Lakshmi Travis, M., & Prasanna Venkatesan, V. (2014). A survey on recent research in business intelligence. *Journal of Enterprise Information Management, 27*(6), 831–866.
2. Atitallah, S. B., Driss, M., Boulila, W., & Ghézala, H. B. (2020). Leveraging deep learning and Iot big data analytics to support the smart cities development: Review and future directions. *Computer Science Review, 38,* 100303.
3. Barenji, A. V., Wang, W., Li, Z., & Guerra-Zubiaga, D. A. (2019). Intelligent E-commerce logistics platform using hybrid agent based approach. *Transportation Research Part E: Logistics and Transportation Review, 126,* 15–31.
4. Burhan, M., Rehman, R. A., Khan, B., & Kim, B.-S. (2018). Iot elements, layered architectures and security issues: A comprehensive survey. *Sensors, 18*(9), 2796.
5. Chen, Y. (2020). Iot, cloud, big data and Ai in interdisciplinary domains. *Simulation Modelling Practice and Theory, 102,* 102070.
6. Desarkar, A., & Das, A. (2017). Big-data analytics, machine learning algorithms and scalable/parallel/distributed algorithms. In *Internet of things and big data Technologies for Next Generation Healthcare* (pp. 159–197). Springer.
7. Dlamini, Z., Francies, F. Z., Hull, R., & Marima, R. (2020). Artificial intelligence (Ai) and big data in cancer and precision oncology. *Computational and Structural Biotechnology Journal, 18,* 2300–2311.
8. Dubey, R., Gunasekaran, A., Childe, S. J., Bryde, D. J., Giannakis, M., Foropon, C., Roubaud, D., & Hazen, B. T. (2020). Big data analytics and artificial intelligence pathway to operational performance under the effects of entrepreneurial orientation and environmental dynamism: A study of manufacturing organisations. *International Journal of Production Economics, 226,* 107599.
9. El-Sappagh, S. H., & El-Masri, S. (2014). A distributed clinical decision support system architecture. *Journal of King Saud University – Computer and Information Sciences, 26*(1), 69–78.
10. Fang, R., Pouyanfar, S., Yang, Y., Chen, S.-C., & Iyengar, S. (2016). Computational health informatics in the big data age: A survey. *ACM Computing Surveys (CSUR), 49*(1), 1–36.
11. Gubbi, J., Buyya, R., Marusic, S., & Palaniswami, M. (2013). Internet of things (Iot): A vision, architectural elements, and future directions. *Future Generation Computer Systems, 29*(7), 1645–1660.
12. Ilmudeen, A. (2021). Design and development of Iot-based decision support system for dengue analysis and prediction: Case study on Sri Lankan context. In *Healthcare paradigms in the internet of things ecosystem* (pp. 363–380). Elsevier.
13. Ilmudeen, A. (2020). Big data, artificial intelligence, and the internet of things in cross-border E-commerce. In *Cross-border E-commerce marketing and management* (pp. 257–272). IGI Global.

14. Kedra, J., & Gossec, L. (2020). Big data and artificial intelligence: Will they change our practice? *Joint, Bone, Spine, 87*(2), 107–109.
15. Kumar, P. M., Lokesh, S., Varatharajan, R., Chandra Babu, G., & Parthasarathy, P. (2018). Cloud and Iot based disease prediction and diagnosis system for healthcare using fuzzy neural classifier. *Future Generation Computer Systems, 86*, 527–534.
16. Kumar, S. R., Gayathri, N., Muthuramalingam, S., Balamurugan, B., Ramesh, C., & Nallakaruppan, M. (2019). Medical big data mining and processing in E-healthcare. In *Internet of things in biomedical engineering* (pp. 323–339). Elsevier.
17. Leung, K., Choy, K. L., Siu, P. K., Ho, G. T., Lam, H., & Lee, C. K. (2018). A B2c E-commerce intelligent system for re-engineering the E-order fulfilment process. *Expert Systems with Applications, 91*, 386–401.
18. Li, X. (2018). Development of intelligent logistics in China. In *Contemporary logistics in China* (pp. 181–204). Springer.
19. Luan, Y., & Zhang, Z. (2018). Research on E-commerce integrated management information system of cross-border enterprises based on collaborative information middleware. *Information Systems and e-Business Management, 18*(4), 527–543.
20. Mahmud, S., Iqbal, R., & Doctor, F. (2016). Cloud enabled data analytics and visualization framework for health-shocks prediction. *Future Generation Computer Systems, 65*, 169–181.
21. Manogaran, G., Lopez, D., Thota, C., Abbas, K. M., Pyne, S., & Sundarasekar, R. (2017). Big data analytics in healthcare internet of things. In *Innovative healthcare systems for the 21st century* (pp. 263–284). Springer.
22. Mayo, C. S., Mierzwa, M., Moran, J. M., Matuszak, M. M., Wilkie, J., Sun, G., Yao, J., Weyburn, G., Anderson, C. J., Owen, D., & Rao, A. (2020). Combination of a big data analytics resource system with an artificial intelligence algorithm to identify clinically actionable radiation dose thresholds for dysphagia in head and neck patients. *Advances in Radiation Oncology, 5*(6), 1296–1304.
23. Plageras, A. P., Psannis, K. E., Stergiou, C., Wang, H., & Gupta, B. B. (2018). Efficient Iot-based sensor big data collection–processing and analysis in smart buildings. *Future Generation Computer Systems, 82*, 349–357.
24. Pramanik, M. I., Lau, R. Y., Demirkan, H., & Azad, M. A. K. (2017). Smart health: Big data enabled health paradigm within smart cities. *Expert Systems with Applications, 87*, 370–383.
25. Ranjan, J., & Foropon, C. (2021). Big data analytics in building the competitive intelligence of organizations. *International Journal of Information Management, 56*, 102231.
26. Ravi, V., & Kamaruddin, S. (2017). Big data analytics enabled smart financial services: Opportunities and challenges. In *Big data analytics* (pp. 15–39). Springer.
27. Subramaniyaswamy, V., Manogaran, G., Logesh, R., Vijayakumar, V., Chilamkurti, N., Malathi, D., & Senthilselvan, N. (2018). An ontology-driven personalized food recommendation in Iot-based healthcare system. *The Journal of Supercomputing, 75*(6), 3184–3216.
28. Syafrudin, M., Alfian, G., Fitriyani, N. L., & Rhee, J. (2018). Performance analysis of Iot-based sensor, big data processing, and machine learning model for real-time monitoring system in automotive manufacturing. *Sensors (Basel), 18*(9), 2946.
29. Tu, Y., & Shangguan, J. Z. (2018). Cross-border E-commerce: A new driver of global trade. In *Emerging issues in global marketing* (pp. 93–117). Springer.
30. Vijayakumar, V., Malathi, D., Subramaniyaswamy, V., Saravanan, P., & Logesh, R. (2019). Fog computing-based intelligent healthcare system for the detection and prevention of mosquito-borne diseases. *Computers in Human Behavior, 100*, 275–285.

Technological Developments in Internet of Things Using Deep Learning

Rakesh Chandra Joshi, Saumya Yadav, and Vibhash Yadav

1 Introduction

The growth of embedded devices interconnected with mobile phones leads to the concept of IoT [1], which results in a sensor-rich world. Objects in our environment are enriched with sensing, computing, and communication capabilities such as fan, lights, watches, etc. Such capabilities promise to transform the interactions between physical objects and humans. Hence, substantial efforts have been made to build user-friendly and smarter applications on embedded devices and mobile phones. At the same time, deep learning keeps growing by developing computing devices based on human-centric data such as audio, image, speech, and video data. Using deep learning with IoT devices could bring advanced applications that can perform object detection, object recognition, and other complex sensing in real time, thus supporting human interaction with physical objects present around them. Before the IoT era, most research in deep learning was to improve the deep neural network algorithms and generate models to work efficiently when the scale of the problem grows to big data, and import models on cloud platforms. The arrival of IoT has then opened up a totally different direction when the scale of the problems shrank down to resource-constrained devices and to the need for real-time analytics. IoT is widely used in different applications such as smart cities, e-health care [2], cybersecurity [3], industrial engineering, and vehicle systems. With the growth of IoT sensor nodes [4], the power consumption has also increased. Therefore, the reduction of power consumption in IoT devices is a crucial challenge for the researchers. Some

R. C. Joshi · S. Yadav (✉)
Centre for Advanced Studies, Dr. A.P.J. Abdul Kalam Technical University, Lucknow, India
e-mail: rakesh@cas.res.in; saumya@cas.res.in

V. Yadav
Department of Information Technology, Rajkiya Engineering College,
Banda, Uttar Pradesh, India

© The Author(s), under exclusive license to Springer Nature Switzerland AG 2022
F. Al-Turjman et al. (eds.), *Transforming Management with AI, Big-Data, and IoT*, https://doi.org/10.1007/978-3-030-86749-2_15

of the main IoT-based application problems are improving power efficiency and network longevity by dynamic motion, limiting battery life, and reducing size of the sensor nodes of IoT. Deep learning-based approaches are a solution for network communication. IoT-based approaches are a solution for network communication, as it can efficiently manage the power consumption and optimize the IoT application. Also, it can be used for monitoring networks, making a prediction on batteries, etc.

This chapter presents an overview on different Internet of Things (IoT)-based techniques with different deep learning techniques. Various such applications with the help of smart devices interconnected to other devices are discussed.

2 Internet of Things (IoT)

The Internet of Things or IoT is the interconnection of physical objects called "things" embedded with an electronic chip, sensors, and other forms of hardware. These objects can be composed of different computing devices with unique identifiers, for example, Radio Frequency Identifier (RFID) tags, but does the task without the human-human or human-computer interaction. Internet-enabled smart devices are used to collect the data, process it, and exchange the data with other systems and devices connected through web connection. These smart objects can communicate with other connected devices that can be monitored and remotely controlled. The number of such devices are increasing day by day, and collected data size is also growing.

The most typical IoT architecture is composed of three layers: perception layer, network layer and application layer. Figure 1 illustrates the different layers of the IoT architecture with associated sensors, devices and applications.

Perception layer or sensor layer is the bottom layer of the architecture, which can connect different sensors with the network. This layer functions with multiple sensors, actuators and RFID devices, which are used to measure, collect, and process the data of state information. This collected information is transmitted to upper layers with different interfaces.

Network layer or transmission layer is the middle layer that processes the data and information received from the perception layer and estimates the transmission

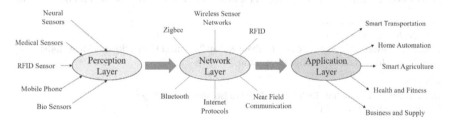

Fig. 1 Three-layer IoT architecture

route with different networks associated with it. It is a crucial layer of the architecture that associates different network devices such as switch, gateway, hub etc. with different communication technologies, that is, Wi-Fi, LTE, Bluetooth, ZigBee, etc. are different communication protocols.

Application layer or business layer is the topmost layer of the IoT architecture that collects the transmitted data from the network layer and utilises the data for requisite operations and services to end devices and networks. It determines the set for protocols at application level for passing the information. It defines various IoT-based applications that can be deployed, for example, smart health, smart grid, smart homes, smart cities, etc.

Apart from the sensor data, multimedia data [5] also becomes crucial when utilizing the IoT devices for the task. Multimedia data can be a video from a surveillance system or fingerprint data in the form of images. Thus, the multimedia data size is also increasing with the increasing number of sensor devices. Therefore, to manage the massive amount of multimedia data and its processing, deep learning can help in a significant manner.

3 Deep Learning

Deep learning [6] is the class of machine learning that employs the multiple layer non-linear information processing system to extract the features for analysis and classification of data. Deep learning tries to imitate the human brain's functioning for data processing and pattern formation to make decisions. It utilizes the artificial neural networks similar to the human brain where neurons are connected with each other and form a layered architecture for processing the data for detection, classification, or any other data-related task without involving manual feature engineering. The raw data is given to the first layer of the neural network, and information is extracted automatically which is passed through multiple layers to train the model and make decision. Deep learning is flexible and has more power and simple use, which makes it suitable to use in different fields such as research, medicine, art, finance, biology, image processing, robotics, natural language processing, and various domains of science. Deep learning allows a system to learn complex mapping

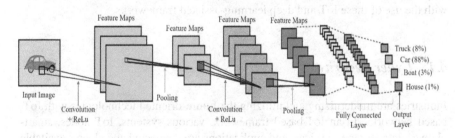

Fig. 2 Deep learning-based multilayer architecture

functions to relate the input data directly with the output without using handcrafted features. The general architecture of deep learning is shown in Fig. 2 for image classification task [7] using different set of training images and multilayer functionality.

Although the concept of artificial neural network and deep learning is old, the last decade has seen a great progress in this field because of three factors:

1. Advancement in the chip processing technology, which results in higher data storage and fast processing of the data, that is, graphic processing units (GPUs)
2. Increase in the size of the data for training the deep learning model, for example, ImageNet, Microsoft COCO, PlantVillage and many more
3. Advancement in deep learning and machine learning approaches for processing the information, for example, deployment of Generative Adversarial Networks (GANs), TensorFlow, Keras, Sklearn, etc.

These factors made possible the research for hierarchical feature representation and exploration of more complex non-linear functions and made use of labelled and unlabelled data for automatic feature extraction.

4 Deep Learning in Internet of Things (IoTs) with Different Applications

IoT is the network of different interconnected devices and sensors where information is passed in a meaningful manner. The addition of deep learning with IoT can make more advantages and more efficient decisive system. Figure 3 illustrates the methodology and working of the IoT-based integration with deep learning.

IoT sensors and associated sensor nodes capture the data in different parameters according to the application. Then, collected data is uploaded to the cloud servers with different gateways, Wi-Fi, Bluetooth or by any other means. The collected data is analysed by the deep learning model which is already trained with the large dataset in given set of parameters. The train model takes the data and classifies the data in different classes to make a decision. The predicted decision is passed to the mobile application associated with different requisite people or points to make the output available in easy manner. Different applications and services can be deployed with the use of these IoT and deep learning-assisted frameworks.

4.1 Cyber Security

Industries are modernizing and utilizing the future-oriented technologies with IoT-based networks. In an IoT-based framework, various systems, IoT nodes, smartphones, data storage, services, and applications are connected and always available

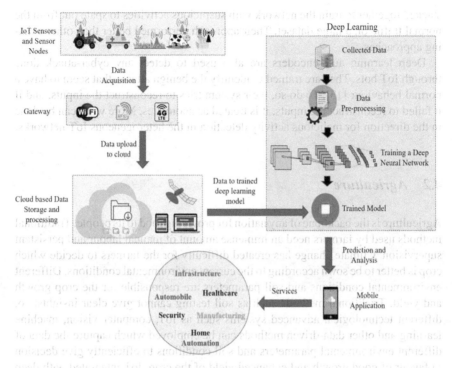

Fig. 3 Integration of deep learning with Internet of Things (IoT)

through the Internet, containing sensitive and important user data. Thus, IoT-based system can be prone to different cyber-attacks with different gateways, for example, security breach, distributed denial-of-service (DDoS) attack, spam, data leakage, etc. The security of IoT is highly compromised by different malware attacks and software piracy which can cause financial loss and loss of prestige of the institution with these kind of losses and data manipulations [8, 9]. Source codes of an individual or organization can be illegally reused to recreate the software using the reverse engineering procedure or the same logic to generate another software or code. Thus, deep learning-based approaches can be used to detect malware-infected files and pirated software for an IoT network. A deep learning-based neural network is proposed in [10] with TensorFlow where weighting feature and tokenization techniques are used for noisy data filtration and to check source code plagiarism. Malware files are converted into colour images, and visualised features are used for malicious infections in the IoT network.

Most of cyber-attacks that have affected IoT networks were DDoS-based attacks. Roopak et al. proposed and tested different deep learning and machine learning models to detect DDoS attacks and found the hybrid CNN-LSTM model performed better than others in CICIDS2017 dataset [11].

Detection of malicious traffic data is done using a stacked deep learning approach for IoT devices in [12]. Five ResNet-based pre-trained deep learning networks are stacked together to train the network with suspicious activities to spate out from the normal traffic on a large dataset. Their approach performed better than other existing approaches.

Deep learning auto-encoders are also used to detect any cyber-attack done through IoT bots. They are trained to identify the benign attacks that seem to have a normal behaviour [13]. To do so, their system tries to reconstruct the inputs, and if it failed to reconstruct the inputs, it is treated as anomalies. More work can be done in the direction for malicious activity detection in the heterogeneous IoT networks.

4.2 Agriculture

Agriculture is the backbone of any nation for providing food to its people. Traditional methods used by farmers need an immense amount of manual labour and persistent supervision. Climate change has created difficulty for the farmers to decide which crop is better to be sown according to the current environmental conditions. Different environmental conditions and soil parameters are responsible for the crop growth and yield. Traditional methods such as soil testing cannot give clear insights; so, different technological advanced systems such as IoT, computer vision, machine learning and other data-driven methods can be employed which capture the data of different environmental parameters and soil conditions to efficiently give decision in favour of good growth and enhanced yield of the crop. IoT integrated with deep learning can give a choice to the farmer for proper selection of suitable crop and improve the irrigation system in the crop field.

The bottom layer in IoT-based smart agriculture is aimed at collecting data in certain time intervals from the various sensors that were deployed to measure the different environment-based parameters such as intensity of light, humidity, carbon dioxide concentration, temperature, and soil moisture. The next step is the pre-processing where collected data is analysed and incomplete data is removed. Next, data compression is done to reduce the size of the data so that more information can be transmitted utilising less bandwidth. Afterwards, different artificial intelligence and deep learning algorithms are used to process the data in cloud-based servers to efficiently analyse the data and retrieve important information to establish the analogy to implement a smart IoT-based agriculture framework. Different deep learning techniques can be used to control different sensors to change the environmental parameters according to which crop can adapt to the current atmosphere and crop growth can be enhanced.

Muangprathub et al. [14] has proposed an IoT-based application for watering the crops. Different wireless sensor nodes are used in IoT network to capture data of different environmental factors and soil moisture, humidity and temperature can be controlled in the crop field with use of data mining. Mobile and web-based applications have been created to control the watering of crop and information of the yield.

Rainfall data is an important parameter for the agriculture. Rainfall forecast is done using deep learning, and weather pattern is analysed from the climate-related data. A hybrid climate learning model is a developed multilayer perceptron network that chooses one from multiple forecast [15]. The hybrid LSTM network and system use chosen forecast to find the relationship of predicted forecast and corresponding rainfall and temperature observations. Finally, the system predicts the next-day rainfall forecast.

Nutritive value of dairy farm forage is of great importance for producers and technicians. But assessment of quality is time-consuming, has high cost, and needs manual work which creates the need of simple, automatic and fast device to get physical and chemical attributes of the dairy farm forage. A portable system is presented in [16] for analysis of nutritional values of dairy farm forage where different near-infrared spectrometry (NIRS) techniques are used. Different IoT-based tools were used to send the collected data to cloud servers to process which can be accessed by any device. Analytical parameters such as concentration of required substance and measured spectrum from NIR spectroscopy are used to establish the relationship with deep learning-based models.

IoT-based devices can be used for multiple applications such as agricultural monitoring [17], soil moisture estimation [18], pH and salinity analysis [19], weather prediction [20], farm machine performance [21], autonomous vehicles and robots for field [22, 23], etc.

4.3 Healthcare

Preventive care and early intervention can be taken for patients at need using distant health monitoring. Nowadays, these kinds of healthcare-assistive systems [24] are readily available with the advent of IoT. Health-related data from pregnant women, elderly people, and patients affected with serious disease can be continuously monitored for analysis. IoT-based healthcare can facilitate more affordable heath assistance outside the hospitals apart from the traditional hospital setting which can help independent living persons. As large amount of data in different formats can be collected from the multi-sensor IoT nodes, they can be easily processed with deep learning-based method in an efficient manner. Multiple models can be trained with the collected dataset, and computation can be done in the cloud with more accuracy in less time (Fig. 4).

Healthcare industry is also seeing more advancement with different technologies. As IoT-based systems provide the resources over the Internet, large amount of health and surveillance data or big data is associated with it. To avoid the latency-related issues, edge computing and fog computing can be done. An ensemble deep learning and IoT-based framework is developed for edge computing devices for analysis of heart diseases and automatic diagnosis [25]. HealthFog efficiently manages the heart patient's data using the IoT devices. Computation time, accuracy, jitter latency, network bandwidth, and power consumption are some of the parameters used to assess the developed system's performance.

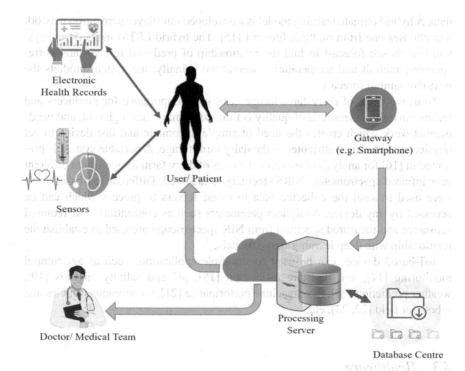

Electronic
Health Records

Sensors

User/ Patient

Gateway
(e.g. Smartphone)

Doctor/ Medical Team

Processing
Server

Database Centre

Fig. 4 Smart healthcare using IoT

Proper intake of nutrient is necessary for the infants, and deficiency of nutrients can create serious problems such as severe disease or failure of organs. These type of health problems can be carried with age. Thus, automatic nutrient monitoring using IoT and deep learning techniques can play an important role to meet these challenges. A fully automatic IoT-based nutrition monitoring system named 'SmartLog' is developed to take care of infants [26]. Bayesian-based five-layer perceptron neural network is developed for analysis of data and its classification.

Cerebral vascular accidents or stroke is a serious medical problem, and its need is to be significantly addressed. An IoT-based framework is proposed to classify the stroke from CT scan images with the use of convolutional neural network. Haemorrhagic stroke and ischemic stroke are differentiated from the normal healthy brain [27]. With the advent of IoT, remote monitoring and diagnosis of strokes can be possible in an efficient manner.

IoT is making a significant advantage in healthcare and makes the system human-independent and less susceptible to human-based errors. More number of the patients can be dealt with the IoT-based system.

4.4 Block Chain

With the development of network, there is a significant growth of data generation in IoT platform as it integrates different devices together [1, 2]. To analyse large amount of data and develop a robust and accurate system for classification, detection, and prediction of objects in IoT, deep learning model is used. It is a robust analytic tool that enables classification and detection of generated IoT data collected from different noisy and complex environments [3]. However, designing an effective deep learning model is a complex task as there can be issues like data poisoning attack, privacy leak of IoT devices [4], lack of useful data for generating some models, single point failure, etc. To overcome this problem, blockchain-based deep learning is introduced to solve the data analysis problem in IoT. Deep learning is used to solve the issue of privacy leak, and blockchain technology can maintain the integrity and confidentiality of deep learning in IoT.

5 Conclusion

IoT and deep learning are widely used in multiple applications throughout the globe. Much research is ongoing towards the further enhancement in the technology for more efficient use and providing services. The effective utilisation of IoT and deep learning technologies in a wide range of applications could result in more information extraction in less time. Performance of traditional sensor-based methods can be enhanced using deep learning. In agricultural applications, farmers can choose suitable crop based on weather and other environmental conditions. Efficient irrigation system in the crop field can be developed using these techniques. IoT-based systems are more prone to the security issues like distributed denial-of-service (DDoS) attack, spam, data leakage, etc. Thus, deep learning models can be trained with data to differentiate between normal conditions and malicious infections in the IoT network. The healthcare sector can be more benefited with the fast diagnosis and real-time monitoring of the patients.

More enhancement can be done to improve the data transmission speed and optimize the received information for fast and real-time decision-making.

References

1. Rathore, S., & Park, J. H. (2018). Semi-supervised learning based distributed attack detection framework for IoT. *Applied Soft Computing, 72*, 79–89.
2. Rathore, S., Sharma, P. K., Sangaiah, A. K., & Park, J. J. (2017). A hesitant fuzzy based security approach for fog and mobile-edge computing. *IEEE Access, 6*, 688–701.

3. Mohammadi, M., Al-Fuqaha, A., Sorour, S., & Guizani, M. (2018). Deep learning for IoT big data and streaming analytics: A survey. *IEEE Communication Surveys and Tutorials, 20*, 2923–2960.
4. Shankar, A., Pandiaraja, P., Sumathi, K., Stephan, T., & Sharma, P. (2020). Privacy preserving E-voting cloud system based on ID based encryption. *Peer-to-Peer Networking and Applications.* https://doi.org/10.1007/s12083-020-00977-4
5. Sharmila, K. D., Kumar, P., & Ashok, A. (2020). Introduction to multimedia big data computing for IoT. In S. Tanwar, S. Tyagi, & N. Kumar (Eds.), *Multimedia big data computing for IoT applications* (Intelligent systems reference library) (Vol. 163). Springer. https://doi.org/10.1007/978-981-13-8759-3_1
6. Almalaq, A., & Zhang, J. J. (2020). Deep learning application: Load forecasting in big data of smart grids. In W. Pedrycz & S. M. Chen (Eds.), *Deep learning: Algorithms and applications* (Studies in computational intelligence) (Vol. 865). Springer. https://doi.org/10.1007/978-3-030-31760-7_4
7. Xin, M., & Wang, Y. (2019). Research on image classification model based on deep convolution neural network. *Journal on Image and Video Processing, 40.* https://doi.org/10.1186/s13640-019-0417-8
8. Zanella, A., et al. (2014). Internet of things for smart cities. *IEEE Internet of Things Journal.* https://doi.org/10.1109/JIOT.2014.2306328
9. Karbab, E. M. B., et al. (2018). MalDozer: Automatic framework for android malware detection using deep learning. In *DFRWS 2018 EU – proceedings of the 5th annual DFRWS Europe.* https://doi.org/10.1016/j.diin.2018.01.007
10. Ullah, F., et al. (2019). Cyber security threats detection in internet of things using deep learning approach. *IEEE Access.* https://doi.org/10.1109/ACCESS.2019.2937347
11. Roopak, M., et al. (2019). Deep learning models for cyber security in IoT networks. In *2019 IEEE 9th annual computing and communication workshop and conference, CCWC 2019.* https://doi.org/10.1109/CCWC.2019.8666588
12. Alotaibi, B., & Alotaibi, M. (2020). A stacked deep learning approach for IoT cyberattack detection. *Journal of Sensors.* https://doi.org/10.1155/2020/8828591
13. Meidan, Y., Bohadana, M., Mathov, Y., et al. (2018). N-BaIoT—Network-based detection of IoT botnet attacks using deep autoencoders. *IEEE Pervasive Computing, 17*(3), 12–22.
14. Muangprathub, J., Boonnam, N., Kajornkasirat, S., Lekbangpong, N., Wanichsombat, A., & Nillaor, P. (2019). IoT and agriculture data analysis for smart farm. *Computers and Electronics in Agriculture, 156,* 467–474.
15. Madhukumar, N., et al. (2020). Consensus forecast of rainfall using hybrid climate learning model. *IEEE Internet of Things Journal,* 1. https://doi.org/10.1109/JIOT.2020.3040736
16. Rego, G., et al. (2020). A portable IoT NIR spectroscopic system to analyze the quality of dairy farm forage. *Computers and Electronics in Agriculture.* https://doi.org/10.1016/j.compag.2020.105578
17. Bauer, J., & Aschenbruck, N. (2018). Design and implementation of an agricultural monitoring system for smart farming. In *2018 IoT vertical and topical summit on agriculture* (p. 1e6). IEEE. https://doi.org/10.1109/IOTTUSCANY.2018.8373022
18. Brinkhoff, J., Hornbuckle, J., Quayle, W., Lurbe, C. B., & Dowling, T. (2017). WiField, an IEEE 802. 11-based agricultural sensor data gathering and logging platform. In *Eleventh international conference on sensing technology (ICST).* IEEE.
19. Popović, T., Latinović, N., Pešić, A., Zečević, Z., Krstajić, B., & Djukanović, S. (2017). Architecting an IoT-enabled platform for precision agriculture and ecological monitoring: A case study. *Computers and Electronics in Agriculture, 140,* 255–265. https://doi.org/10.1016/j.compag.2017.06.008
20. Yan, M., Liu, P., Zhao, R., Liu, L., Chen, W., Yu, X., et al. (2018). Field microclimate monitoring system based on wireless sensor network. *Journal of Intelligent Fuzzy Systems, 35*(2), 1325e1337. https://doi.org/10.3233/JIFS-169676

21. Oksanen, T., Piirainen, P., & Seilonen, I. (2015). Remote access of ISO 11783 process data by using OPC unified architecture technology. *Computers and Electronics in Agriculture, 117*, 141–148. https://doi.org/10.1016/j.compag.2015.08.002
22. Bechar, A., & Vigneault, C. (2016). Agricultural robots for field operations: Concepts and components. *Biosystems Engineering, 149*, 94–111. https://doi.org/10.1016/j.biosystemseng.2016.06.014
23. Christiansen, P., Nielsen, L. N., Steen, K. A., Jørgensen, R. N., & Karstoft, H. (2016). DeepAnomaly: Combining background subtraction and deep learning for detecting obstacles and anomalies in an agricultural field. *Sensors, 16*(1904), 1–21. https://doi.org/10.3390/s16111904
24. Kim, Y. B., Yoo, S. K., & Kim, D. (2006). Ubiquitous healthcare: Technology and service. In N. Ichalkaranje, A. Ichalkaranje, & L. Jain (Eds.), *Intelligent paradigms for assistive and preventive healthcare* (Studies in computational intelligence) (Vol. 19). Springer. https://doi.org/10.1007/11418337_1
25. Tuli, S., Basumatary, N., Gill, S. S., et al. (2019). HealthFog: An ensemble deep learning based smart healthcare system for automatic diagnosis of heart diseases in integrated IoT and fog computing environments. *Future Generation Computer Systems*. https://doi.org/10.1016/j.future.2019.10.043
26. Sundaravadivel, P., et al. (2018). Smart-log: A deep-learning based automated nutrition monitoring system in the IoT. *IEEE Transactions on Consumer Electronics*. https://doi.org/10.1109/TCE.2018.2867802
27. Dourado, C. M. J. M., et al. (2019). Deep learning IoT system for online stroke detection in skull computed tomography images. *Computer Networks*. https://doi.org/10.1016/j.comnet.2019.01.019

Machine Learning Models for Sentiment Analysis of Tweets: Comparisons and Evaluations

Leeladhar Koti Reddy Vanga, Adarsh Kumar, Kamalpreet Kaur, Manmeet Singh, Vlado Stankovski, and Sukhpal Singh Gill

1   Introduction

Sentiment analysis is categorized under text analysis for studying feelings and opinions of people on particular topics. Other points of view have to be considered while making a decision by people, and that contributes towards becoming an imperative procedure for organisations [1]. This method is beneficial for not only people but also for organisations who are keen in understanding the view of other persons. Commonly, to know about people's opinions, we require various ways including

L. K. R. Vanga · S. S. Gill (✉)
School of Electronic Engineering and Computer Science, Queen Mary University of London, London, UK
e-mail: ec19469@qmul.ac.uk; s.s.gill@qmul.ac.uk

A. Kumar
Department of Systemics, School of Computer Science, University of Petroleum and Energy Studies, Dehradun, India
e-mail: adarsh.kumar@ddn.upes.ac.in

K. Kaur
QualiteSoft, Chandigarh, India
e-mail: kamalpreet.kaur@qualitesoft.com

M. Singh
Centre for Climate Change Research, Indian Institute of Tropical Meteorology (IITM), Pune, India

Interdisciplinary Programme (IDP) in Climate Studies, Indian Institute of Technology (IIT), Bombay, India
e-mail: manmeet.cat@tropmet.res.in

V. Stankovski
Faculty of Computer and Information Science, University of Ljubljana, Ljubljana, Slovenia
e-mail: vlado.stankovski@fri.uni-lj.si

surveys or questionnaires or simply ask friends and family members [2]. Generally, aforementioned means are applied if organisations require feedback for services or products. Nevertheless, unpredictable advancement in number of websites and mobile applications result in exponential growth of sustainable content [3]. In recent times, it has become easy for people to share their opinions related to anything on blogs as well as social media.

Sentiment analysis is also considered as a way of collecting viewpoints for discovering subjective text as well as extracting valuable information from particular textual data [2]. Text could be categorised into three categories: as negative or neutral double, positive however, sometimes be neutral. Technology and methods are ways of getting reference information [4]. Statistical models could be used to combine huge amount of textual data, which further supplies the information in supporting organizational decisions. The important mining tool is emotional analysis to offer aid to solve the problems and challenges such as web data analysis and collection of data systematically without any disruption. Further, this data can be disintegrated and distributed for interpretation and analysis [5]. Moreover, specific emotional tone can be achieved using Natural language processing (NLP) using different features such as operator, unit allocation and part explanation [3]. To implement this, there is need to do initial handiwork for law writing, which will help to attain required accuracy in terms of conceptual context [6].

1.1 Motivation and Our Contributions

There are many companies and individual researchers trying to achieve the best results and produce state-of-art models for sentiment/emotional analysis [7]. However, many of the researchers do not compare different types of machine learning algorithms [8] to find a suitable algorithm for a specific problem of sentiment analysis of tweets. There is a need to consider and compare different types of machine learning models and choose the best for a particular task as sentiment analysis is a wide topic [9].

The *main contributions* of this research work are:

1. In this chapter, the two machine learning algorithms are compared to evaluate their performance based on F-1 score.
2. This work has conducted machine-learning-based experimentation and identified the best algorithm for sentiment analysis of tweets.
3. In experimentation, this work presented the data tokenization and stemming process, conducted token stitching, shows positive and negative hashtags, applied feature engineering, and shows the word frequency tables, and applies regression model classification over tweets.
4. Implementation and integration of Term Frequency-Inverse Document Frequency (TF-IDF) increases the F1-score which in turn increases the accuracy from

51.41% (with Decision Tree model) to 54.98%. To further increase the accuracy score, logistic model is used which shows accuracy of 57.72%.

5. Further experimentation is performed to integrate logistic model with TF-IDF and this increased the accuracy to 58.62%. Thus, a comparative analysis of different models in hash tag analysis is performed that identified the best approach in terms of F1-score.

1.2   Chapter Structure

The rest of the chapter is structured as follows: Section 2 presents the impact of social media. Section 3 presents the related work. Section 4 presents the model design. Section 5 describes the implementation and validation of experimental results. Section 6 presents conclusions. Section 7 presents the future work.

2   Impact of Social Media

The best growing region of Natural Language Processing (NLP) research is the sentiment analysis which is performed effortlessly with the usage of Python. The significance of social media, Twitter, microblogs, blogs, forums and emotional analysis has been increased by the growth of social media platforms that include web critique, which has skyrocketed the demand [10]. The accurate investigation of large amount of self-assertive data can help to make effective decisions for business [11]. Further, Artificial Neural Networks (ANN) can be used to perform real-time emotion classification [4]. To perform social media analytics, Twitter is one of the important platforms to collect data from famous celebrities, politicians and people; Twitter helps to create 15 billion API calls and 3 billion tweets everyday [12]. On the basis of many materials, Twitter has a good supply and demand [13]. Further, a number of applications are increasingly related to Twitter, which helps to increase the data collection. Further, this collected data can be used for surveillance systems for future predictions about natural disasters.

3   Related Work

In the literature, various emotional search engines are developed, which use classical queries to find results on any topic [14]. The main use of these engines is to find out the polarity of text documents and classify the retrieved data in three different classes such as neutral, negative or positive. Further, inappropriate languages or nuances can be searched using monitor online forums-based programs [15]. Moreover, other works such as Yue et al. [15] inspected the character of emotions in online communication, which can be helpful to study the emergence and perspectives of psychology [16]. The use of Naive Bayesian Classification (NBC)-based guided learning models (Andy Bromberg) can increase the training speed for future search [17]. NBC is domain-specific, which can be used on trained domain for data distribution, and it achieves classification accuracy less than 70% [18]. Furthermore, Yoo et al. [3] also verified whether the emotional vocabulary is used to find out the tweets which contain negative or positive words. Islam and Zibran [19] used SentiStrength algorithm to develop a database to allocate definite text at double polarity (negative or positive), and SentiStrength marked 298 positive and 465 negative terms. Palominoe et al. [11] identified that SentiStrength produces better results as compares to machine learning-based MySpace model in terms of classification of negative emotions. Pagolu et al. [20] proposed a new algorithm for the support of SentiStrength which increases the search of emotional words from 693 to 2310. Nguyen and Shirai [2] compared various machine learning algorithms for tweet data and found that Logistic Regression is better than SentiStrength. Ceron et al. [21] studied the minds of Twitter users and their actions based on hierarchical party ranking technique. Further, these studies have classified the Twitter data in three categories such as positive, negative or neutral [22–24].

Shaukat et al. [5] have conducted opinion mining from movie reviews, and it has been observed that the use of neural network on the "movie review database" has shown long lists of positive and negative words. These lists can be used to rate the likeness of movies. The proposed system has managed to achieve a very high accuracy (91%), which is found to be a remarkable performance in this area. Additionally, the proposed system has performed quantitative and qualitative insights on different facets of movie parameters. In consideration of these parameters, the proposed system is able to find the negative connotations in positive words. Mathapati et al. [7] has proposed a semi-supervised domain adaptive dual sentiment analysis. This is a trained classifier with labelled data. It has been observed that the accuracy of results can be improved with long-term dependency analysis between the words. Further, a collaborative deep learning approach has been integrated for dual sentimental analysis which in turn is used for sequence prediction and classification of reviews. The use of long short-term memory recurrent neural network has resulted in the reduction of training time and improved the proposed system results. Haselmayer et al. [8] proposed a procedure for fine-grained sentiment score-based analysis. Here, negative and positive sentiment dictionary analysis is performed with resource-intensive hand coding. The experimental results analysis shows that the proposed crowd-based dictionary provides efficient and valid performance measurements. This illustrates the use of tonality of party statements and presents the media reports. The negative sentiment dictionary procedure includes sampling sentences, sentiment strength analysis, tonality score measurements, and discriminating the important and unimportant words. Eldefrawi et al. [9] proposed sentiment analysis of Arabic opinions and performed comparative analysis. This work has proposed sentiment analysis techniques with three elements. These elements include the type of comparative keywords, opinion feature analysis, and entity's position. Here, the proposed technique has considered the limits of human interference to preparing the lexicons, and collecting and categorizing comparative keywords. This work has considered the elements and techniques that do not consider features as well. Results have shown 96.5% F-measure for correctly identified sentiment analysis. Thus, the proposed sentiment-based technique is useful in finding negative and positive opinions.

3.1 Critical Analysis

Table 1 shows the comparison of proposed work with existing works [25, 26]. All the research work have presented sentiment analysis of Twitter data using a single prepossessing step and machine learning technique [25, 26]. There is a need to compare different types of machine learning algorithms to achieve optimal result [12]. As every technique has its own advantages and disadvantages, our proposed work addresses these issues by applying different data pre-processing steps and using different machine learning approaches to sentiment analysis.

4 Design and Methodology

This section explains the design and methodology followed in experimentation. Attitude analysis is the next step after getting access to huge number of sources of data [23]. In this work, we consider only words which are written in English language only. Natural Language Toolkit (NLTK) is used in this work to develop Python programs to play with human data, and it helps to interpret the emotional data. Further, we used two machine learning methods (Linear Regression and Decision Trees) for performance comparison to find out the best technique for the classification of unknown emotions, and the probabilities of negative and positive tweets are distributed [24].

Figure 1 describes the steps for process followed for sentiment analysis. In the proposed steps, machine learning approach is applied where data is passed through pre-processing stages including tokenisation, and stemming which in turn make the data ready to apply multiple machine learning approach with and without use of TF-IDF. Thereafter, the results are compared to identify the best approach and

Table 1 Comparative feature analysis of proposed work with state of the art

Work	Model description				Performance parameter
	Data pre-processing steps		Machine learning algorithms		
	Tokenisation	Stemming	Logistic Regression	Decision Trees	F-1 score
[25]	✓	✓			
[26]	✓				
This work	✓	✓	✓	✓	✓

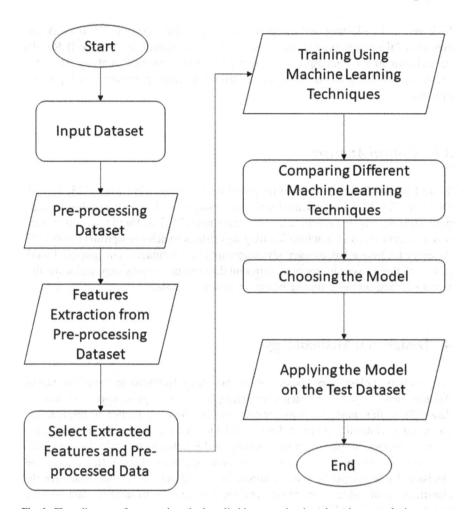

Fig. 1 Flow diagram of proposed method applied in regression-based sentiment analysis

extend it with tweets analysis. The detailed steps involved in the proposed work are explained as follows:

4.1 Dataset

The dataset used in this experiment is available at Kaggle [27]. The training set consists of sample tweets which are associated with labels. Label "0" denotes the tweet is not racist/sexist and label "1" denotes the tweet is racist/sexist. More details about dataset is presented in Sect. 5.2.

4.2 Feature Extraction Techniques

This subsection explains briefly the feature extraction and machine learning techniques applied in tweets analysis and statistics generation of this work.

4.2.1 Bag of Words

Text data can also contain sentiments with inherent feeling or emotions. Bag of Words (BoW) was introduced to recognise the consequence of polarization on distinct words. This will help to check vocabulary and create measures for the existence of entered text in these words. BoW analyses the connection between the attributes and relates the tags within the text data, but local context or simulation is not considered in BoW. Basically, it removes all the words existing in the exercise set but our proposed work does not remove stop words before exercise. Further, it can remove the words used frequently if the entry is missing regularly. Furthermore, it also considers the entry of those words which occur rarely. The occurrence of words is called word frequency, and rare occurrence of words is called conversion rate.

4.2.2 TF-IDF

Term Frequency and Inverse Document Frequency (TF-IDF) is defined as the estimation metric to evaluate the meaning of words for documents or collections statistically. Due to the increase in number of items of the word within the document, it increases the importance, but it is augmented by the occurrence of the word that exists in the text. The formula for TF (Term Frequency) is given in Eq. 1.

$$TF = \frac{\text{Number of times a word (term) appears in a document}}{\text{Total number of words in the document}} \tag{1}$$

The formula for IDF (Inverse Document Frequency) is given in Eq. 2.

$$IDF = \log\left(\frac{\text{Total No. of documents}}{\text{No. of documents with word in it}}\right) \tag{2}$$

4.3 Machine Learning Models

This section explains the machine learning approaches that are applied in experimentation for comparative analysis [12].

4.3.1 Logistic Regression

This is supervised machine learning model and used to solve problems of binary classification. Further, this model helps to forecast the chances of instance related to specific class. Moreover, reviews can be negative or positive, and then this model could be represented as occurrence of positive words.

4.3.2 Decision Tree

It is a tree like machine learning model to perform both classification and regression. In tree-like mode, there are two main entities named as nodes and leaves. Leaves are the places where decisions are made or final outcomes are measured. On the other hand, nodes are places where logical split is made to continue with experimentation and concrete result's observations.

4.4 Performance Evaluation Metric

The proposed implementation uses F1-score in statistics analysis. F1-score is a harmonic mean of Precision and Recall, shown in Eq. 3 below.

$$F1\,Score = 2 * \frac{Precision * Recall}{Precision + Recall} \quad (3)$$

5 Performance Evaluation

This section discusses experimental setup, implementation details and experimental results. More details are presented as follows.

5.1 Configuration Settings

To implement the sentimental analysis, Python version 3.8.5 is used, and experiments are conducted on a system with an Intel®Core™i9-9880HProcessor (16M Cache, 2.3 GHz), 16 GB RAM and 512 GB of SSD running on Mac OS.

5.2 Dataset Overview

The process starts by reading and making a backup copy of the train.csv (training dataset) and test.csv (test dataset) files. An overview of these files is in Figs. 2 and 3. Figure 2 shows that the parameters considered in training dataset include processes identification, label, tweet details and associated processes identification. The training dataset includes 31,962 entities. As training dataset has three attributes id, label and tweet, the label gives us the information whether the tweet is positive or negative. As shown in Fig. 2, if the label is "1" it indicates that the corresponding tweet is of negative sentiment and if the label shows "0" the tweet is of positive sentiment.

Figure 3 shows test dataset. This dataset has two attributes id and tweets. The dataset is unlabelled because as we use this dataset for the final model. The total number of records in this dataset is 17,197. As it can be clearly seen, the entities in this dataset are totally different, and it does not contain any tokenization in its raw state. The tokenization and other processing techniques are performed in pre-processing stages.

5.3 Data Pre-Processing

To prepare dataset for training on the machine learning model, both train and test datasets should be combined. This is done using a function called "dataframe. append" from "pandaspackage". Figure 4 shows the combined results of train and test datasets. In combined results, tweets are combined as well. The test dataset does not contain the label column, and hence the new cells in label column are assigned a "NaN" value as shown in Fig. 4.

```
          id  label                                                tweet
0          1      0    @user when a father is dysfunctional and is s...
1          2      0    @user @user thanks for #lyft credit i can't us...
2          3      0                               bihday your majesty
3          4      0    #model    i love u take with u all the time in ...
4          5      0           factsguide: society now    #motivation
...       ...    ...                                              ...
31957  31958      0    ate @user isz that youuu?ð      ð     ð     ð     ð       ð...
31958  31959      0       to see nina turner on the airwaves trying to...
31959  31960      0    listening to sad songs on a monday morning otw...
31960  31961      1    @user #sikh #temple vandalised in in #calgary,...
31961  31962      0                     thank you @user for you follow

[31962 rows x 3 columns]
```

Fig. 2 An overview of training dataset

```
         id                                                          tweet
0       31963   #studiolife #aislife #requires #passion #dedic...
1       31964    @user #white #supremacists want everyone to s...
2       31965   safe ways to heal your #acne!!      #altwaystohe...
3       31966   is the hp and the cursed child book up for res...
4       31967    3rd #bihday to my amazing, hilarious #nephew...

...       ...                                                          ...
17192   49155   thought factory: left-right polarisation! #tru...
17193   49156   feeling like a mermaid ð        #hairflip #neverre...
17194   49157   #hillary #campaigned today in #ohio((omg)) &am...
17195   49158   happy, at work conference: right mindset leads...
17196   49159   my   song "so glad" free download!  #shoegaze ...


[17197 rows x 2 columns]
```

Fig. 3 An overview of test dataset

```
     id  label                                                  tweet
0    1    0.0   @user when a father is dysfunctional and is s...
1    2    0.0  @user @user thanks for #lyft credit i can't us...
2    3    0.0                              bihday your majesty
3    4    0.0  #model   i love u take with u all the time in ...
4    5    0.0             factsguide: society now     #motivation
5    6    0.0  [2/2] huge fan fare and big talking before the...
6    7    0.0   @user camping tomorrow @user @user @user @use...
         id  label                                                  tweet
49152   49153   NaN   we love the pretty, happy and fresh you! #teen...
49153   49154   NaN   2_damn_tuff-ruff_muff__techno_city-(ng005)-web...
49154   49155   NaN   thought factory: left-right polarisation! #tru...
49155   49156   NaN   feeling like a mermaid ð        #hairflip #neverre...
49156   49157   NaN   #hillary #campaigned today in #ohio((omg)) &am...
49157   49158   NaN   happy, at work conference: right mindset leads...
49158   49159   NaN   my   song "so glad" free download!  #shoegaze ...
```

Fig. 4 An overview of combined train and test datasets

The next step in data pre-processing includes removing twitter handles "@User" shown in Fig. 5 as this does not contribute anything that helps the model to improve the accuracy of classification.

The third step includes removal of all the numbers, special characters and punctuation, a sample of dataset after removing all these can be seen in Fig. 6. Like the twitter handles these also do not contribute the model performance.

The next step is removal of short words. This process is critical as the length of the words to remove affects our model performance. In this experiment, the length

```
     id  ...                                              Clean_Tweets
0    1   ...        when a father is dysfunctional and is so sel...
1    2   ...        thanks for #lyft credit i can't use cause th...
2    3   ...                                     bihday your majesty
3    4   ...  #model   i love u take with u all the time in ...
4    5   ...              factsguide: society now     #motivation
5    6   ...  [2/2] huge fan fare and big talking before the...
6    7   ...                   camping tomorrow          dannyâ€¦

[7 rows x 4 columns]
```

Fig. 5 Dataset after removing twitter handles

```
     id  ...                                              Clean_Tweets
0    1   ...        when a father is dysfunctional and is so sel...
1    2   ...        thanks for #lyft credit i can t use cause th...
2    3   ...                                     bihday your majesty
3    4   ...  #model   i love u take with u all the time in ...
4    5   ...              factsguide  society now     #motivation
5    6   ...        huge fan fare and big talking before the...
6    7   ...                   camping tomorrow          danny

[7 rows x 4 columns]
```

Fig. 6 Dataset after removing all numbers, special characters and punctuation

of stop words is 3. This length is randomly selected. The experimentation is easy to be modified with other length parameters. Figure 7 shows the sample dataset after all the three-letter words are removed.

Figure 8 describes the last step of data pre-processing: tokenisation and stemming. Tokenisation helps to convert the tweets into individual tokens for the next step stemming. NLTK package is used for stemming which helps to remove the suffixes for words, that is, all the "ing", "ly", "es", "s". Now, the refined data is almost ready for further experimentation, as can be easily understood from Fig. 8.

Finally, the tokens are put back together for training our model as shown in Fig. 9.

```
     id  ...                                        Clean_Tweets
 0    1  ...  when father dysfunctional selfish drags kids i...
 1    2  ...  thanks #lyft credit cause they offer wheelchai...
 2    3  ...                               bihday your majesty
 3    4  ...                            #model love take with time
 4    5  ...                     factsguide society #motivation
 5    6  ...  huge fare talking before they leave chaos disp...
 6    7  ...                            camping tomorrow danny

[7 rows x 4 columns]
```

Fig. 7 Dataset after removing all the stop words

```
0    [when, father, dysfunctional, selfish, drags, ...
1    [thanks, #lyft, credit, cause, they, offer, wh...
2                            [bihday, your, majesty]
3                    [#model, love, take, with, time]
4               [factsguide, society, #motivation]
Name: Clean_Tweets, dtype: object
0    [when, father, dysfunct, selfish, drag, kid, i...
1    [thank, #lyft, credit, caus, they, offer, whee...
2                            [bihday, your, majesti]
3                    [#model, love, take, with, time]
4                   [factsguid, societi, #motiv]
Name: Clean_Tweets, dtype: object
```

Fig. 8 Dataset after tokenisation and stemming

5.4 Data Visualisation

This is one of the most important steps in machine learning projects as it helps us to get an idea about the dataset. In this experiment, a visualisation technique called WordCloud is used to visualise the most frequent words proportional to the size, as shown in Fig. 10 and Fig. 11. The importance of this experimentation is to clearly reflect the importance of words. Larger the size of words means more frequently it occurs in the dataset. Thus, the importance of those words increases accordingly.

All the hashtags are extracted from the tweets. These include hashtags from positive tweets and negative tweets. Further, these hashtags are represented in a frequency scale. This helps to visualise how many numbers of times a hashtag is used.

```
      id   ...                                         Clean_Tweets
0     1    ...    when father dysfunct selfish drag kid into dys...
1     2    ...    thank #lyft credit caus they offer wheelchair ...
2     3    ...                               bihday your majesti
3     4    ...                          #model love take with time
4     5    ...                            factsguid societi #motiv
5     6    ...    huge fare talk befor they leav chao disput whe...
6     7    ...                                 camp tomorrow danni

[7 rows x 4 columns]
```

Fig. 9 Dataset after putting stitching the tokens

Fig. 10 Visual representation of words with label "0"

Figures 12 and 13 show the 17 most used hash tags for both positive and negative tweets.

5.5   Feature Engineering

This subsection explains more feature-based analysis of results. More details are presented as follows.

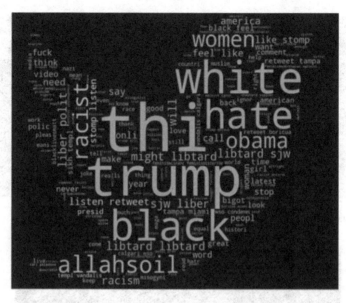

Fig. 11 Visual representation of words with label "1"

Fig. 12 17 Most used
positive hashtags

	Hashtags	Count
32	love	1654
39	posit	917
58	smile	676
157	healthi	573
38	thank	534
176	fun	463
343	life	425
92	affirm	423
286	summer	390
4	model	375
355	cute	367
400	beauti	365
315	happi	358
20	blog	356
223	friend	343
455	fathersday	341
22	gold	301

Fig. 13 17 most used
negative hashtags

	Hashtags	Count
22	trump	136
31	polit	95
63	allahsoil	92
30	liber	81
28	libtard	77
29	sjw	75
60	retweet	63
59	black	46
62	miami	46
33	hate	37
70	bigot	36
61	tampa	32
257	obama	32
182	blm	28
250	hispan	28
20	brexit	27
174	sikh	27

5.5.1   Bag of Words

As the name says "bag of words", a large set of words are analysed in this experimentation that lies in one package. With the help of these features, this work has
created a frequency table of all the words as shown in Fig. 14. Figure 14 shows that
there are 49,159 rows in the table, and frequency table has 49,158 entities represented with identification numbers. The column values corresponding to entities
represent frequencies.

5.5.2   TF-IDF

The next process applies TF-IDF. With the help of this feature on the dataset, this
work obtained the TF-IDF table of the whole dataset, as shown in Fig. 15.

	0	1	2	3	4	5	6	...	993	994	995	996	997	998	999
0	0	0	0	0	0	0	0	...	0	0	0	0	0	0	0
1	0	0	0	0	0	0	0	...	0	0	0	0	0	0	0
2	0	0	0	0	0	0	0	...	0	0	0	0	0	0	0
3	0	0	0	0	0	0	0	...	0	0	0	0	0	0	0
4	0	0	0	0	0	0	0	...	0	0	0	0	0	0	0
...	...	...	...	...	...	...	...	...	...	...	...	...	...	...	...
49154	0	0	0	0	0	0	0	...	0	0	0	0	0	0	0
49155	0	0	0	0	0	0	0	...	0	0	0	0	0	0	0
49156	0	0	0	0	0	0	0	...	0	0	0	0	0	0	0
49157	0	0	0	0	0	0	0	...	0	0	0	0	0	0	0
49158	0	0	0	0	0	0	0	...	0	0	0	0	0	0	0

[49159 rows x 1000 columns]

Fig. 14 Frequency table of all the words in the dataset

	0	1	2	3	4	5	6	...	993	994	995	996	997	998	999
0	0.0	0.0	0.0	0.0	0.0	0.0	0.0	...	0.0	0.0	0.0	0.0	0.0	0.0	0.0
1	0.0	0.0	0.0	0.0	0.0	0.0	0.0	...	0.0	0.0	0.0	0.0	0.0	0.0	0.0
2	0.0	0.0	0.0	0.0	0.0	0.0	0.0	...	0.0	0.0	0.0	0.0	0.0	0.0	0.0
3	0.0	0.0	0.0	0.0	0.0	0.0	0.0	...	0.0	0.0	0.0	0.0	0.0	0.0	0.0
4	0.0	0.0	0.0	0.0	0.0	0.0	0.0	...	0.0	0.0	0.0	0.0	0.0	0.0	0.0
...	...	...	...	...	...	...	...	...	...	...	...	...	...	...	...
49154	0.0	0.0	0.0	0.0	0.0	0.0	0.0	...	0.0	0.0	0.0	0.0	0.0	0.0	0.0
49155	0.0	0.0	0.0	0.0	0.0	0.0	0.0	...	0.0	0.0	0.0	0.0	0.0	0.0	0.0
49156	0.0	0.0	0.0	0.0	0.0	0.0	0.0	...	0.0	0.0	0.0	0.0	0.0	0.0	0.0
49157	0.0	0.0	0.0	0.0	0.0	0.0	0.0	...	0.0	0.0	0.0	0.0	0.0	0.0	0.0
49158	0.0	0.0	0.0	0.0	0.0	0.0	0.0	...	0.0	0.0	0.0	0.0	0.0	0.0	0.0

[49159 rows x 1000 columns]

Fig. 15 TF-IDF table for the dataset

6 Experimental Results and Performance Comparison

In first experimentation, this work has used Logistic Regression model. First, Bag-of-Words feature is applied and probability of the tweet being positive or negative is calculated. Then, resulting value is converted into F1-score. F1-score of logistic model using Bag-of-Words feature is 0.572135. This means the model can predict with an accuracy of 57.72%.

Next, the same logistic model is trained with TF-IDF feature.

This gave an F1-score of 0.586206. This model has an accuracy of 58.62%.Now, Decision Tree is used for second model. Like Logistic Regression this work used Bag-of-Words and TF-IDF for this model. It gave a score of 0.514177 and 0.549882, respectively; this means Decision Tree model that has used Bag-of-Words feature predicts with 51.41% accuracy and the one that used TF-IDF predicts with 54.98% accuracy.

We comparing both Logistic Regression and Decision Trees models with their F1-scores. Table 2 shows the F1-scores of the both models with Bag-of-Words as feature. Results shows that Logistic Regression approach is having better F1-score compared to Decision Tree.

Table 3 shows the F1-scores of both models with TF-IDF as feature. According to this comparative analysis, Logistic Regression approach is better than Decision Tree approach for a given set of data.

After comparative analysis of outputs of the models (taken in experimentation), it has been decided to continue further experimentation with the use of Logistic Regression model with TF-IDF as feature on the test dataset, as shown in Fig. 16.

After analysing the results shown in Fig. 16, tweet predictions followed by labelling can be easily observed. For example, Id corresponds to a particular tweet and label "0" tells us if the tweet is positive and label "1" denotes negative tweet.

7 Conclusions

This work compared two machine learning techniques for sentiment analysis of tweets. The main aim is to analyse large amount of pre-labelled twitter dataset using different data pre-processing steps, features extraction techniques and machine learning algorithms to help predict the sentiment behind the tweets. Logistic Regression model that was subjected to TF-IDF feature gave an accuracy rate of 58.62% over Decision Tree which gave an accuracy of 54.98%. Therefore, this model is considered to classify the tweets.

7.1 Future Directions

Apart from the models discussed, it has been observed that there are many other machine learning models that can be used for sentiment/emotional analysis. These models go beyond positive-negative classification.

Our future challenge includes exploration of different machine learning models to predict discrete emotion like happiness, sadness etc. the proposed work can be compared with similar machine learning models such as Support Vector Machine (SVM), K-Nearest Neighbours (KNN), Naive Bayes, Random Forest and Ensemble

Table 2 F1-scores of models using Bag-of-Words as feature

Performance parameter	Machine Learning Model	
	Logistic Regression	Decision Trees
F1-score	0.572135	0.514178

Table 3 F1-scores of models using TF-IDF as feature

Performance	Machine Learning Model	
parameter	Logistic Regression	Decision Trees
F1-score	0.586207	0.549882

Fig. 16 Logistic Regression model classifying tweets as positive or negative

```
              id  label
0           31963      0
1           31964      0
2           31965      0
3           31966      0
4           31967      0

...          ...    ...
17192       49155      1
17193       49156      0
17194       49157      0
17195       49158      0
17196       49159      0

[17197 rows x 2 columns]
```

Voting, and Bidirectional Encoder Representations from Transformers (BERT) model and graph machine learning model using different performance parameters such as Precision and Recall, Specificity and Sensitivity.

Further, data collected using various sensors in IoT network is found to be helpful because this will meet the requirements of futuristic applications. Thus, sentimental analysis in IoT application needs to be explored in future [12]. There is a possibility of implementing sentiment analysis using deep learning strategy and natural language processing. Such strategies can be integrated with distributed ensemble models in Edge and Fog computing environments [12]. Furthermore, such algorithms can also be used to do real-time large-scale mental health analysis which is crucial for keeping population in check in extreme circumstances like in COVID-19 pandemic [28, 29].

References

1. Dhaoui, C., Webster, C. M., & Tan, L. P. (2017). Social media sentiment analysis: Lexicon versus machine learning. *Journal of Consumer Marketing, 34,* 480–488.
2. Nguyen, T. H., Shirai, K., & Velcin, J. (2015). Sentiment analysis on social media for stock movement prediction. *Expert Systems with Applications, 42,* 9603–9611.

3. Yoo, S., Song, J., & Jeong, O. (2018). Social media contents based sentiment analysis and prediction system. *Expert Systems with Applications, 105*, 102–111.
4. Nguyen, T. H., & Shirai, K. (2015). Topic modeling based sentiment analysis on social media for stock market prediction. In *Proceedings of the 53rd Annual Meeting of the Association for Computational Linguistics and the 7th International Joint Conference on Natural Language Processing. 1*, 1354–1364.
5. Shaukat, Z., Zulfiqar, A. A., Xiao, C., Azeem, M., & Mahmood, T. (2020). Sentiment analysis on IMDB using lexicon and neural networks. *SN Applied Sciences, 2*, 1–10.
6. Gaspar, R., Pedro, C., Panagiotopoulos, P., & Seibt, B. (2016). Beyond positive or negative: Qualitative sentiment analysis of social media reactions to unexpected stressful events. *Computers in Human Behavior, 56*, 179–191.
7. Mathapati, S., Nafeesa, A., Tanuja, R., Manjula, S. H., & Venugopal, K. R. (2019). Semi-supervised domain adaptation and collaborative deep learning for dual sentiment analysis. *SN Applied Sciences, 1*, 907.
8. Haselmayer, M., & Jenny, M. (2017). Sentiment analysis of political communication: Combining a dictionary approach with crowdcoding. *Quality and Quantity, 51*, 2623–2646.
9. Eldefrawi, M. M., Elzanfaly, D. S., Farhan, M. S., & Eldin, A. S. (2019). Sentiment analysis of Arabic comparative opinions. *SN Applied Sciences, 1*, 411.
10. Vilares, D., Thelwall, M., & Alonso, M. A. (2015). The megaphone of the people? Spanish SentiStrength for real-time analysis of political tweets. *Journal of Information Science, 41*, 799–813.
11. Palomino, M., Taylor, T., Göker, A., Isaacs, J., & Warber, S. (2016). The online dissemination of nature–health concepts: Lessons from sentiment analysis of social media relating to "nature-deficit disorder". *International Journal of Environmental Research and Public Health, 13*, 142.
12. Gill, S. S., Tuli, S., Xu, M., Singh, I., Singh, K. V., Lindsay, D., et al. (2019). Transformative effects of IoT, blockchain and artificial intelligence on cloud computing: Evolution, vision, trends and open challenges. *Internet of Things, 8*, 100118.
13. Wang, Y., & Li, B. (2015). Sentiment analysis for social media images. In *IEEE International Conference on Data Mining Workshop (ICDMW)*, 1584–1591.
14. Muhammad, A., Wiratunga, N., & Lothian, R. (2016). Contextual sentiment analysis for social media genres. *Knowledge-Based Systems, 108*, 92–101.
15. Yue, L., Chen, W., Li, X., Zuo, W., & Yin, M. (2019). A survey of sentiment analysis in social media. *Knowledge and Information Systems, 60*, 1–47.
16. Vashishtha, S., & Susan, S. (2019). Fuzzy rule based unsupervised sentiment analysis from social media posts. *Expert Systems with Applications, 138*, 112834.
17. Etter, M., Colleoni, E., Illia, L., Meggiorin, K., & D'Eugenio, A. (2018). Measuring organizational legitimacy in social media: Assessing citizens' judgments with sentiment analysis. *Business & Society, 57*, 60–97.
18. Beigi, G., Hu, X., Maciejewski, R., & Liu, H. (2016). An overview of sentiment analysis in social media and its applications in disaster relief. In W. Pedrycz & S. M. Chen (Eds.), *Sentiment analysis and ontology engineering* (Vol. 639, pp. 313–340). Springer.
19. Islam, M. R., & Zibran, M. F. (2018). SentiStrength-SE: Exploiting domain specificity for improved sentiment analysis in software engineering text. *Journal of Systems and Software, 145*, 125–146.
20. Pagolu, V. S., Reddy, K. N., Panda, G., & Majhi, B. (2016). Sentiment analysis of Twitter data for predicting stock market movements. In *IEEE international conference on signal processing, communication, power and embedded system (SCOPES)*, 1345–1350.
21. Ceron, A., Curini, L., & Iacus, S. M. (2016). iSA: A fast, scalable and accurate algorithm for sentiment analysis of social media content. *Information Sciences, 367*, 105–124.
22. Xia, R., Jiang, J., & He, H. (2017). Distantly supervised lifelong learning for large-scale social media sentiment analysis. *IEEE Transactions on Affective Computing, 8*, 480–491.

23. Younis, E. M. (2015). Sentiment analysis and text mining for social media microblogs using open source tools: An empirical study. *International Journal of Computer Applications, 112*, 44–48.
24. Thakor, P., & Sasi, S. (2015). Ontology-based sentiment analysis process for social media content. In *INNS Conference on Big Data*, 199–207.
25. Agarwal, A., Xie, B., Vovsha, I., Rambow, O., & Passonneau, R. J. (2011). Sentiment analysis of twitter data. In *Proceedings of the workshop on language in social media*, 30–38.
26. Sahayak, V., Shete, V., & Pathan, A. (2015). Sentiment analysis on twitter data. *International Journal of Innovative Research in Advanced Engineering (IJIRAE), 2*, 178–183.
27. Twitter sentiment analysis, Online Available: https://www.kaggle.com/arkhoshghalb/twitter-sentiment-analysis-hatred-speech. Accessed 16 10 2020.
28. Tuli, S., Tuli, S., Tuli, R., & Gill, S. S. (2020). Predicting the growth and trend of COVID-19 pandemic using machine learning and cloud computing. *Internet of Things*, 100222.
29. Kumar, A., Sharma, K., Singh, H., Naugriya, S. G., Gill, S. S., & Buyya, R. (2021). A drone-based networked system and methods for combating coronavirus disease (COVID-19) pandemic. *Future Generation Computer Systems, 115*, 1–19.

Secure and Enhanced Crowdfunding Solution Using Blockchain Technology

Lakshit Madaan, Dikshita Jindal, Amit Kumar, Suresh Kumar, and Mahaveer Singh Naruka

1   Introduction

Crowdfunding is a mechanism in which some capital is required from a large number of individuals to invest in some business ventures. It can be used through social websites by bringing together many entrepreneurs and investors together, with the potential to increase entrepreneurship and expand the pool of investors and venture capitalists [2]. Crowdfunding has created various opportunities for many entrepreneurs. They raised millions of dollars by deploying their projects to various crowdfunding websites such as Kickstarter and Indiegogo. These websites attract thousands of people to interact with the projects and invest in them so that in return they could also receive benefits through them. According to some surveys, crowdfunding is mostly synonymous with Kickstarter as it is the largest crowdfunding platform. Two types of users drive Kickstarter.

L. Madaan · D. Jindal
Department of Computer Science and Engineering, HMR Institute of Technology and
Management, Delhi, India

A. Kumar (✉)
School of Engineering and Technology, Sharda University, Uttar Pradesh, India
e-mail: amit.kumar@hmritm.ac.in

S. Kumar
Department of computer science and engineering, NSUT East Campus Formerly AIACTR,
Delhi, India
e-mail: sureshkumar@aiactr.ac.in

M. S. Naruka
Nodal Officer-Academics, State Project Implementation Unit, Uttar Pradesh, India

1. *Creators*: Creators are the ones who create innovative projects and deploy them for funding. They also create pages where they can list information and also add some videos, photos, etc. of their projects so that the backers can get proper details about the projects, and they also set a goal for funding and the deadline, plus rewards where backers can achieve on some contribution to the project.
2. *Backers*: Backers are the ones who fund the project by donating some amount to them so that the project can reach its goal for funding and backers also in return are rewarded according to their contribution [3].

According to some surveys in 2010, over one-and-a-half billion dollars have been transferred from Kickstarter backer to project creator, and Kickstarter 5% cut with each dollar means the revenue collected was around 75–80 billion dollars which is a huge amount. Still, the most important question that arises is the transparency of Kickstarter [4]. Many fraud and failure cases are also reported, such as a product CST-01 (Central Standard Timing) 'the world's thinnest watch', which raised more than 1 million dollars but could not keep up to their promise. Thus, the technology which is best and trending to solve the issue of transparency in crowdfunding is 'blockchain'; blockchain is a distributed ledger and is mainly known for its transparency and can be used as an even more legitimate way for funding a vast spectrum of projects and causes. Blockchain is basically a digital ledger of transactions that is distributed among various networks on the blockchain [5]; each block on blockchain contains the hash value of the previous block, and its hash value is like a chain of blocks combined to form blockchain, and if someone tries to tamper with them, the hash values of each block changes. The key feature of blockchain is to provide transparency and trust; every transaction made on the blockchain can be viewed and cannot be deleted. Thus, we can say blockchain is the answer to the question raised earlier about the issues faced due to transparency and fraud in crowdfunding.

2 Blockchain

Blockchain [19] is an immutable distributed ledger, making the history of any digital asset unalterable and transparent using cryptographic algorithms and decentralization [4]. Blockchain [22] is a simple way of passing any type of information from A to B or transaction etc. in a decentralized manner and maintaining transparency [5]. The term blockchain [1] is only heard with cryptocurrency, but apart from this, this technology is used in various other fields such as Banking system, Supply chain, Crowdfunding, Land Registry, Voting, Storing data such as passports, ID cards etc.

Given below is diagram 1 which shows the working of blockchain. The first block of the blockchain is called a genesis block and contains a hash value of 0000; it can be also called block 0 of blockchain. Block 1 contains the previous hash value of genesis block. As we can see that all the blocks are interconnected to each other according to their hash values, each block contains a hash value of the previous block.

When a transaction takes place, a new block is added to the blockchain, it is first mined and verified by the miners; miners solve the given problem and the one who solves faster gets to add the block to the blockchain and also receives some awards like in Ethereum miners receive 3 ether for mining a block. Each block contains a public and a private key; they are generated using cryptographic algorithms like ECC (Elliptic-Curve Cryptography), which is used in Ethereum because it uses 256 bits as compared to RSA which also uses 3072 bits [6]. Using shorter key can result in less computational power and a fast and secure connection. A block contains three different values:

1. *Data:* Data can be any information sent from one user to another.
2. *Hash Value*: Hash value is the current hash value of the block which is generated when the block is added; they are generated using hashing algorithms such as SHA-1 or Keccak-256, etc.
3. *Previous Hash Value*: Previous hash value is the hash value of the previous block.

Suppose someone tries to tamper with blockchain, the hash value of the block changes and as we know all the blocks are connected. In that case, the hash values of all the blocks will change. Still, there are some attacks such DDOS (distributed denial-of-service) attacks and the 51% attacks in which miners or group of miners control 50% of a network's mining power, hash rate, and computing power to mine new blocks [7]. Attackers uses 51% of attacks to reverse the transactions that have taken place, also known as double spending. Thus, there are some challenges to the technology such as 51% attacks and setting up the environment for blockchain, hiring developers, etc., which is quite expensive right now. Still, these challenges would not remain like this in the upcoming future (Fig. 1).

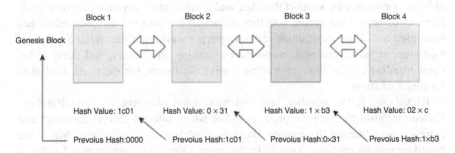

Fig. 1 Hash connection diagram for Blockchain

3 Literature Survey

Zibin Zeng et al. explain a survey blockchain in their paper where they compare the different aspects of applications [20] such as sensor networks [21] and challenges [26] faced by the blockchain technology and mining management ways to increase throughput and decrease processing time. Cynthia Weiyi Cai et al. explain and identify gaps in economics and finance and research how crowdfunding and blockchain can be used in the finance industry. Hasnan Baber et al. explain about how crowdfunding is beneficial in the current scenarios and also introduce a model 'WHIRL' which is a blockchain-based crowdfunding model [23] that assures the members to get their project funded. Waheeda Dhokley et al. explain about blockchain crowdfunding model and introduce a blockchain-based model system 'CROWDSF' which integrates both crowdfunding and blockchain to achieve security. Zhao Hongjiang et al. explain the problems faced in crowdfunding like investor abuse, illegal transactions, frauds, etc. and propose a solution using blockchain to overcome these issues.

4 Conventional Vs Blockchain Crowdfunding

4.1 Conventional Crowdfunding

In simple terms, crowdfunding means that many people come together to invest in small or large amounts to fund a project related to any business. Crowdfunding expands the scope of investing online via a variety of social media platforms. Online funding helps in the huge engagement of various investors and entrepreneurs across the world to invest in the projects [8]. Even small amounts of funding from many people help generate a great amount to start working on the project finally. This modern crowdfunding model is generally based on three types of actors: the project initiator who proposes the idea or project to be funded, individuals or groups who support the idea, and a moderating organization (the 'platform') that brings the parties together to launch the idea [9]. Crowdfunding has been used to finance substantial varietal entrepreneurial outlines related to travel, medicine, airlines, transport, e-commerce, fashion and beauty, and many alike. Crowdfunding can be done on various online platforms, but the most used platform is Kickstarter.

Kickstarter is the number one and most used platform for crowdfunding. Kickstarter aims to bring the projects to life where interest groups, investors and entrepreneurs finance the various projects available on its platform, which are raised by various creators. Kickstarter has been a successful platform in funding a lot of projects. It is an excellent platform if one wants to bring their creative ideas to life on a budget because people adhere to such budget projects as they do not want to risk too much. Many projects on Kickstarter such as EOS, Filecoin, Star

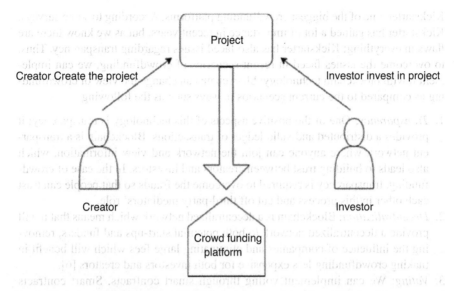

Fig. 2 Work flow diagram for Crowdfunding platform

Citizen, Tezos were able to raise funds over $10 million. This is merely because these projects might be innovative, and people certainly would benefit from it. Although, an out-of-the-box idea with innovation and creativity is always eye-catching. However, there have been many frauds on Kickstarter as well. In the real world, the word of honour may not be maintained and ultimately leads to the destruction of fraudulent companies' dreamy promises. One such example is 'Beef jerky made from Kobe Beef' where this campaign raised more than $120,000 in funding, which was 50 times more than its actual funding goal [10]. On investigation, it was found out that it was all fake including the tasters of the beef. Even Kickstarter had no information about the organization behind it. Such frauds generally leave the backers who fund these campaigns with disappointment and definitely with empty pockets. Many other examples are also there where organizations cannot fulfil the promises made by them to investors. Given below is a block diagram of conventional crowdfunding platforms. There are two types of parties that play a significant role: one is a creator, and the other is an investor [11]. The creator creates the project, and investors invest in particular projects, and they are linked to a crowdfunding platform (Fig. 2).

4.2 Blockchain Crowdfunding

Blockchain is a peer-to-peer network; it is a distributed ledger that can be implemented in various scenarios such as crowdfunding. Crowdfunding is a process in which investors invest in products listed on crowdfunding platforms such as

Kickstarter, one of the biggest crowdfunding platforms. According to some surveys, Kickstarter has gained a lot of importance in recent years, but as we know there are flaws in everything; Kickstarter has also faced issues regarding transparency. Thus, to overcome the issues faced in recent scenarios of crowdfunding, we can implement it with blockchain technology; blockchain can change the view of crowdfunding as compared to the current scenarios in ways such as the following:

1. *Transparency:* One of the positive aspects of this technology is transparency; it provides a distributed and valid ledger of transactions. Blockchain is a transparent network where anyone can join the network and view information, which also leads to building trust between creators and investors. In the case of crowdfunding, transparency is required to overcome the frauds so that people can trust each other in this process and cut off third-party mediators' role.
2. *Decentralization*: Blockchain is a decentralized network which means that it will provide a decentralized network to both potential start-ups and funders, removing the influence of companies and eliminating large fees which will benefit in making crowdfunding less expensive for both investors and creators [6].
3. *Voting*: We can implement voting through smart contracts. Smart contracts are used for building business logic on the blockchain [25]; they are lines of codes that can be executed when terms and conditions are met [7]. We can use smart contracts to implement voting in the crowdfunding process which will benefit the investors so that they can vote for a request relating to investments made by the creator of the project, and those votes can be recorded on the blockchain.
4. *Opportunities:* Many opportunities can be given to both creator and investors as blockchain crowdfunding will provide a decentralized network where the role of small companies and third-party mediators will reduce which will result in eliminating large fees and will make the process less expensive; equal opportunities can be given to both investors and creators by providing them public access and full control on the network [8].

Given below is a block diagram of the Blockchain crowdfunding platform [24] in which there is a creator who creates the project and deploys it on the platform, and investors can donate to a particular project and can become part of the project; apart from all this blockchain voting is used in which creator can request for funding and investors can vote to whether approve the request or not, if the request is approved after voting then the request can be initiated further, else it will be rejected by the investor (Fig. 3).

Given below is a table (Fig. 4) which highlights the key point differences between conventional crowdfunding and blockchain crowdfunding.

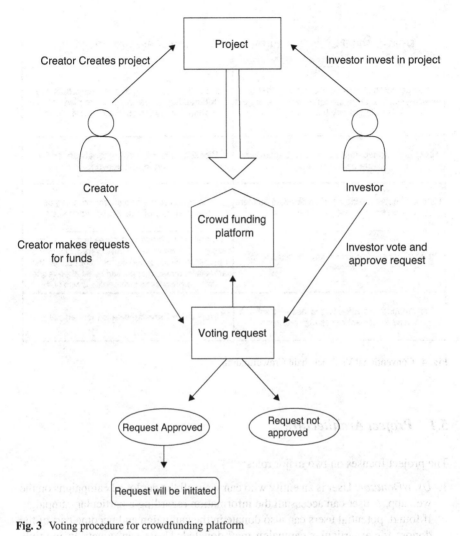

Fig. 3 Voting procedure for crowdfunding platform

5 Result and Analysis

In this section, the authors have proposed a model to change the current scenario process and improve the ongoing difficulties in crowdfunding. The model's primary purpose is that we can use blockchain in crowdfunding and make the process transparent.

Conventional Crowdfunding	Blockchain Crowdfunding
A large number of people come together to invest in small or large amounts to fund a project related to any business.	Blockchain is a peer to peer network, it is a distributed ledger which can be implemented in various scenarios such as crowdfunding.
There is no transparency as to how transactions are being made.	Blockchain is a transparent network and a distributed network.
It is not a secure method of crowdfunding as there is no transparency to the donors.	It is a secure method of crowdfunding as transparency of data flow is maintained
It does not provide voting opportunities to the donors.	It provides voting opportunities to donors. For example, the manager of the contract can make a request for funding and donors can vote on that particular request. If approved by all, then request will be processed further else it won't be.
Conventional crowdfunding is less efficient as many fraudulent cases are reported.	Blockchain Crowdfunding is more efficient.

Fig. 4 Conventional Vs Blockchain Crowdfunding

5.1 Project Architecture

The project focuses on two major roles:

1. *User/Donator:* User is an entity who can view all the deployed campaigns on the web app. A user can access all the information regarding a particular campaign. If found, potential users can also donate in the campaign and become part of the donors for a particular campaign they donated. Users can donate in multiple campaigns and can access all the perks given to a donor for a particular campaign. Suppose a user donated 2 ether and in return the campaign gives the user a 5% share in the company, just like these different perks can be given to a donor in return for donating. Another interesting feature given to the user is voting the donors for a particular vote on a request initiated by the creator of the campaign. Request made can be any type but primarily it is regarding the further investments like a creator can make a request to buy some hardware from a vendor; he will add the request and donors can approve or reject them., if the approved creator can finalize the request, and the amount will be deducted from the campaign contribution. The amount will be sent directly to the vendor; thus, the issue of fraud will be resolved here as instead of giving the amount to the creator it is been transferred directly to the vendor [12].

2. *Creator:* Creator can create a campaign by giving some minimum contribution, which is in Wei (the smallest unit of ether, 1 Ether = 1000000000000000000 Wei). The campaign gets added on the web page with its address; donors can view the campaign. The creator can request as per the requirement for building a product and the vendor's address, donors can reject or approve the project after approving the creator can only finalize the request.

5.2 Modules

The authors have used different modules in the project. These modules are as follows:

5.2.1 Minimum Contribution

Creators can contribute a minimum amount in Wei to deploy the campaign on the web app. Without giving a minimum contribution, the creator can deploy its project.

5.2.2 Single Campaign

Donor or creator can view the single campaign information; this module shows the information regarding a particular campaign, information such as manager or creator address, requests, donors or approvers of the campaign amount. And it also gives the donor the option to donate in the campaign and can view all the requests made or finalized.

5.2.3 Request Module

Creator can make requests for a particular item regarding building a project similar to the creator's request buying a battery. Creator will fill a form with fields such as description, the amount required and the recipient address; the recipient is the vendor from which the battery will be purchased. The request will be added and the donors can view, approve, or reject the request. After voting, if approved, the creator will finalize the request and the amount will be deducted from the campaign account and sent to the recipient directly.

5.3 Technology

The authors have used Ethereum for backend and React and Next.js for frontend Given below is the system architecture of our proposed system (Fig. 5):

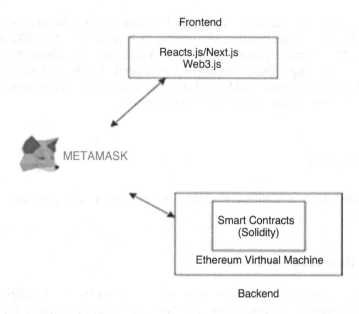

Fig. 5 System Architecture for proposed system

5.3.1   Frontend

For the client side we have used React.js and Next.js; React.js is a library used for building frontend web apps, Next.js is a library on React.js which is used for server-side rendering and web3.js is JavaScript library which is used for interacting with Ethereum node using HTTP.

5.3.2   Metamask

Metamask is a cryptocurrency wallet and is used as a chrome extension. It is used as an interface for interacting with Ethereum blockchain. Metamask is used for storing Ethereum accounts; it is secure and efficient to use and we can also deploy to main Ethereum networks using Metamask. Our project is deployed to Rinkeby test network.

5.3.3   Ethereum

Ethereum is best known for cryptocurrency but it can also be used for business logic means we can use Ethereum for building smart contracts. Smart contracts are the layer on the blockchain which are used for building business logic on the blockchain; they are programs that govern the behaviour of accounts with Ethereum state [13]. We use Solidity language for building smart contracts; it is a high-level object-oriented language which is influenced by C++, Python, and JavaScript to target Ethereum virtual machines (EVMs).

Why Is Ethereum Used?

Ethereum is best suited for building Dapps because of the following:

1. *Solidity*: Ethereum has its own very high-level and object-oriented language, that is, Solidity; it provides a human-readable format which can also be understood by machines, and it is also a combination of JavaScript, Python, and C++.
2. *Hashing technique:* Hashing is a process of converting strings into random values using mathematical functions, and it is one way to enable security [16] during the transactions in the blockchain. Ethereum uses Keccak-256 which produces 256 bits hash. Keccak is a versatile cryptographic algorithm used for hashing and a winner of a multi-year contest held by NIST. It provides a sufficient hashing entropy for a proof-of-work system and is best suited with Dapps. Keccak256() can also be used as a hashing function in solidity [14].
3. *Public-key cryptography:* The use of public-key cryptography in the blockchain maintains security. Public keys are widely used and private keys are kept hidden; so while encoding a message, we use public keys, and while decoding it we use private keys. Ethereum uses ECDSA (Elliptic Curve Digital Signature Algorithm) for public key cryptography which uses public/private key pairs means for every public key there is a private key. We can share our public key with anyone because deriving a private key from a public key is next to impossible. Another type of cryptographic algorithm is RSA, but it is not efficient as ECDSA because RSA uses 3072 public keys for computational complexity, and ECDSA uses only 256 public keys for computational complexity and ECDSA has the same security [17] as RSA but uses less bits that's why it is more efficient than RSA.

ECDSA is based on equation $y^2 = (x^3 + a*x + b)$ mod p [9]. This is based on the equation of a curve on a graph where 'y' is squared, and for any X coordinate, we have two values of y and the curve is symmetric to x axis. The modulo will have only prime numbers and are in range of 160 bits; the modulo 'p' means the possible values of y^2 are between 0 to $p-1$ which gives us 'p' possible values. However, we are dealing with integers; it gives us 'N' possible points on the curve and $N < p$. Since 'x' is having 2 possible points which means $N/2$ possible 'x' valid coordinates on the curve. The graph given below shows the equation (Fig. 6).

The curve from our graph is based on equation $y^2 = (x^3 + a * x + b)$ mod p where $(a = -4, b = 0)$ which is symmetric to x-axis and the points P, Q, and R means if we add point P to Q or draw a line from P to Q it will intersect point R and will lead to Point S where R is equal to negative S ($R = -S$) to represent point R on x-axis [10].

ECDSA signature algorithm requires this equation for verifying and creating a signature, $y^2 = (x^3 + a * x + b)$mod p where a, b are parameters, p is prime modulo and N is number of points on the curve, but now we add a point 'G' needed for ECDSA and it is known as a 'reference point'. For creating a private key, we can use 160 bits, and for public keys we use a point on curve generated via multiplication with G with a private key. Suppose we set 'dA' (some random no.) as private key and 'QA' (a point on curve) as public key so the formula will be $QA = dA * G$.

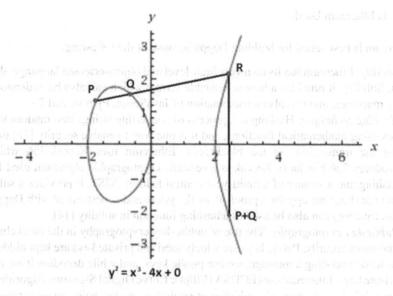

Fig. 6 Derived equation graph for proposed system.

Now, for creating a signature, a signature of 40 bits is required where 20–20 bits each, the first one is called R and the other S, the pair of (R, S) constitutes a signature; to calculate two values we must generate random number k (20 bits) and use point multiplication to calculate point P = k*G, the point's x value is 'R' and for calculating S, we need a SHA1 hash of the message which gives a 20 byte value which will be represented by 'z'. We can calculate S using equation S = k^−1 (z + dA * R) mod p [11] where k^−1 represents inverse of 'k' and (k^−1 * k)mod p = 1. k is a random number used for generating R, z is hash of message and dA is private key, R is x coordinate of P = k * G.

As we have created the signature successfully, so we can verify it using equation P = S^−1 * z * G + S^−1 * R * Qa; if x coordinate of point P is R means signature is valid else not. To verify this, we will substitute P = S^−1 * z * G + S^−1 * R *Q a, Therefore, Qa = dA * G, So P = S^−1 * z * G + S^−1*R*dA*G. Take S^−1 and G common, P = S^−1 * G(z + R*dA) x coordinate of P must match R and R is x coordinate of k * G.

So P = k*G, k * G = S^−1 * G(z + R*dA). Cancelling out G both sides, K = S^−1 * (z + R*dA).

By inverting k and S, we get S = k^−1(z + R*dA); this is the equation used for creating signature; if signature matches, then it is valid.

4. *Gas fee:* Gas fees are used for paying the energy and the computational powers to set and get the functions in the Ethereum blockchain [15]. Ethereum blockchain uses a concept of gas to measure the amount of energy used for a particular smart contract. Gas is measured in wei, which is the smallest unit of ether; 1 ether = 10^18 wei.

5.4 Result

The screenshots of our proposed system are given below with features explained (Fig. 7).

This is the landing page of our project which shows the trending campaigns in the form of a card component with the address given of a particular campaign, and we can also view the campaign details by clicking the 'view campaign' link. In the top right, we have a button from where we can make or add a campaign, and some perks of donation are also given below (Fig. 8).

With the use of this component, we can add a campaign by filling the particular details asked like Minimum contribution in Wei, campaign name, owner name, etc. (Fig. 9)

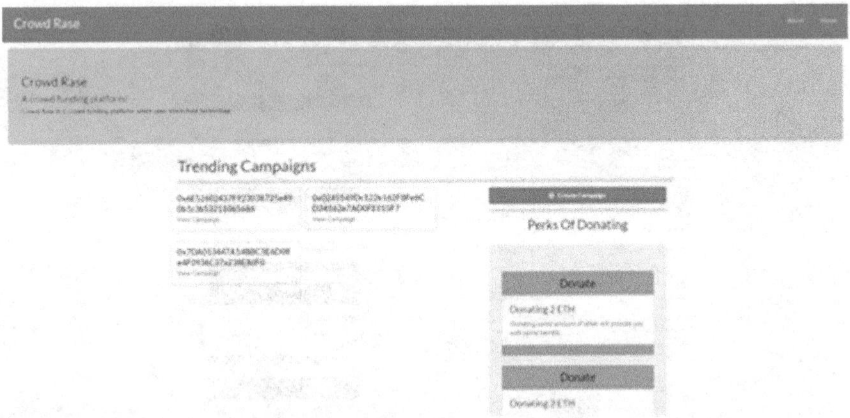

Fig. 7 Landing page including trending campaigns

Fig. 8 Create campaign page of proposed system

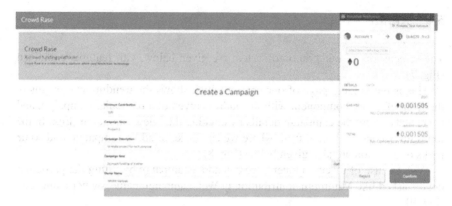

Fig. 9 Metamask account status page for proposed system

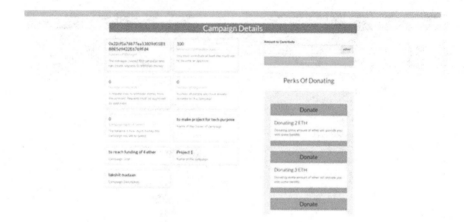

Fig. 10 Campaign details page for proposed system

On submitting, we get a pop of our Metamask account which shows the gas details used for calling this function and the total ethers used. Confirming it takes around 20 seconds, and we get redirected to the home page where our campaign gets added (Fig. 10).

We can click on the 'view campaign' link and get redirected to the campaign detail page. Here we can look for the campaign details like owner name, manager address, number of approvers and requests made to the campaign, etc. (Fig. 11)

If any user likes a particular campaign, they can contribute to it and can be a part of the approver's list. After becoming an approver, he or she can participate in voting the request and can also take advantage of perks of donation. On submitting, we get a pop-up from Metamask account for the transaction as we can see the user is donating 2 ether plus the gas fees (Fig. 12).

The owner or the manager of the particular campaign can make a request for funding by filling the amount, description of the funding, and the recipient to whom we have to send the money. The manager has to pay a certain amount of gas fees to call the function and make a request (Fig. 13).

Fig. 11 Contribution page for proposed system

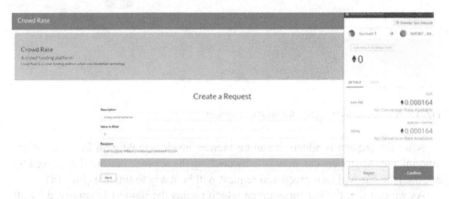

Fig. 12 Funding request cration page for proposed system

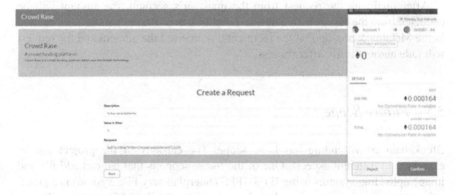

Fig. 13 Request status and description page for proposed system

Fig. 14 Request approval status page for proposed system

Fig. 15 Amount deduction page for proposed system

After the request is added, it can be viewed inside a table with Id, description, amount, etc. Approvers can now vote by clicking the approve button; if approved by all, then the bar will turn green and request will be ready to finalize (Fig. 14).

As we can see, the bar turns green which means the request is approved by all and is ready to finalize by the manager. The manager can only finalize the request from its account else the transaction will fail (Fig. 15).

After finalizing the request from the manager's account, the amount will be deducted from the campaign balance and will be sent to the recipient. As we can see in the Metamask pop-up that the recipient has received the amount and the request will fade automatically after success.

5.5 Future Scope

Blockchain crowdfunding has large scope. The capacity of the project can be expanded to different aspects. One of the future scope is that we can add file and image uploading systems using IPFS. IPFS (Interplanetary File System) is a protocol for peer-to-peer network and to share data in a distributed file system. We can store the hash of a particular image or a file and we can store it on the distributed network. Moreover, the more utilitarian projects can be updated accordingly on the trending list based on the number of people investing in that particular project. We can also take this project on the next level by creating an authentication system for particular individuals like creators and investors to manage their profiles and donate in various campaigns.

6 Conclusion

Crowdfunding with blockchain has a substantial future. The development of such technology can do economical wonders in various financial fields. Our chapter aims to highlight the key features and differences between conventional crowdfunding and blockchain crowdfunding. We have proposed a model in our chapter that focuses on how we can use blockchain in crowdfunding. The model overcomes all the drawbacks found in conventional crowdfunding such as transparency and security. Blockchain is best known for its decentralized structure and transparency; the transactions made using blockchain are pellucid and reliable. Thus, we can see all the transactions on the distributed ledger which makes blockchain crowdfunding overcome fraudulent and transparency issues. There are many challenges faced by blockchain technology like setting up this technology is quite expensive; however, apart from these challenges, this technology has many positive aspects, which we have covered in our model, making it unique.

References

1. Nakamoto, S. Bitcoin: A peer-to-peer electronic cash system. Self-published Paper, 2008 [Online]. Available: https://bitcoin.org/bitcoin.pdf
2. Zheng, Z., Xie, S., Dai, H.-N., Chen, X., & Wang, H. (2018). Blockchain challenges and opportunities: A survey. *International Journal of Web and Grid Services, 14*(4), 352–375.
3. Saini, H., Bhushan, B., Arora, A., & Kaur, A. (2019). Security vulnerabilities in Information communication technology: Blockchain to the rescue (A survey on Blockchain Technology). 2019 2nd International Conference on Intelligent Computing, Instrumentation and Control Technologies (ICICICT). https://doi.org/10.1109/icicict46008.2019.8993229
4. Bentov, I., Gabizon, A., & Mizrahi, A. (2016). Cryptocurrencies without proof of work. In *International conference on financial cryptography and data security* (pp. 142–157). Springer.
5. Kiayias, A., Russell, A., David, B., & Oliynykov, R. (2017). Ouroboros: A provably secure proof-of-stake blockchain protocol. In *Annual international cryptology conference* (pp. 357–388). Springer.
6. Wood, G. (2014). Ethereum: A secure decentralised generalised transaction ledger. *Ethereum project yellow paper, 151*, 1–32.
7. Arora, D., Gautham, S., Gupta, H., & Bhushan, B. (2019). Blockchain-based security solutions to preserve data privacy and integrity. 2019 international conference on computing, communication, and intelligent systems (ICCCIS). https://doi.org/10.1109/icccis48478.2019.8974503
8. Croman, K., Decker, C., Eyal, I., Gencer, A. E., Juels, A., Kosba, A., Miller, A., Saxena, P., Shi, E., Sirer, E. G., et al. (2016). On scaling decentralized blockchains. In *International conference on financial cryptography and data security* (pp. 106–125). Springer.

9. Luu, L., Narayanan, V., Zheng, C., Baweja, K., Gilbert, S., & Saxena, P. (2016). A secure sharding protocol for open blockchains. In *Proceedings of the 2016 ACM SIGSAC conference on computer and communications security* (pp. 17–30). ACM.

10. Kokoris-Kogias, E., Jovanovic, P., Gasser, L., Gailly, N., & Ford, B. (2017). Omniledger: A secure, scale-out, decentralized ledger. *IACR Cryptology ePrint Archive, 2017*, 406.

11. Sharma, T., Satija, S., & Bhushan, B. (2019). Unifying blockchain and IoT: Security requirements, challenges, applications and future trends. 2019 international conference on computing, communication, and intelligent systems (ICCCIS). https://doi.org/10.1109/icccis48478.2019.8974552

12. Liu, T., Li, J., Shu, F., Wu, Y., & Han, Z., et al. (2018). Incentive mechanism design for two-layer wireless edge caching networks using contract theory. *IEEE Transactions on Services Computing*.

13. Liu, T., Li, J., Shu, F., Tao, M., Chen, W., & Zhu, H. (2017). Design of contract-based trading mechanism for a small-cell caching system. *IEEE Transactions on Wireless Communications, 16*(10), 6602–6617.

14. Liu, X., Wang, W., Niyato, D., Zhao, N., & Wang, P. (2018). Evolutionary game for mining pool selection in blockchain networks. *IEEE Wireless Communications Letters*.

15. Sinha, P., Rai, A. K., & Bhushan, B. (2019). Information Security threats and attacks with conceivable counteraction. 2019 2nd international conference on intelligent computing, instrumentation and control technologies (ICICICT). https://doi.org/10.1109/icicict46008.2019.8993384

16. Bhardwaj, A., Al-Turjman, F., Kumar, M., Stephan, T., & Mostarda, L. (2020). Capturing-the-invisible (CTI): Behavior-based attacks recognition in IoT-oriented industrial control systems. *IEEE Access*, 1–1. https://doi.org/10.1109/ACCESS.2020.2998983

17. Shankar, A., Pandiaraja, P., Sumathi, K., Stephan, T., & Sharma, P. (2020). Privacy preserving E-voting cloud system based on ID based encryption. *Peer-to-Peer Networking and Applications*. https://doi.org/10.1007/s12083-020-00977-4

18. Jan, M. A., et al. (2021). Security and blockchain convergence with Internet of Multimedia Things: Current trends, research challenges and future directions. *Journal of Network and Computer Applications, 175*, 102918. https://doi.org/10.1016/j.jnca.2020.102918

19. Yadav, S. P., Mahato, D. P., & Linh, N. T. D. (2020). *Distributed artificial intelligence: A modern approach* (1st ed.). CRC Press. https://doi.org/10.1201/9781003038467

20. Yadav, S. P. (2020). Vision-based detection, tracking and classification of vehicles. *IEIE Transactions on Smart Processing and Computing, SCOPUS, ISSN: 2287-5255, 9*(6), 427–434. https://doi.org/10.5573/IEIESPC.2020.9.6.427

21. Stephan, T., Al-Turjman, F., Suresh Joseph, K., & Balusamy, B. (2020). Energy and spectrum aware unequal clustering with deep learning based primary user classification in cognitive radio sensor networks. *International Journal of Machine Learning and Cybernetics*. https://doi.org/10.1007/s13042-020-01154-y

22. Yadav, S. P., Agrawal, K. K., Bhati, B. S., et al. (2020). Blockchain-based cryptocurrency regulation: An overview. *Computational Economics*. https://doi.org/10.1007/s10614-020-10050-0

23. Baber, H. (2020). Blockchain-based crowdfunding. In R. Rosa Righi, A. Alberti, & M. Singh (Eds.), *Blockchain technology for industry 4.0. Blockchain technologies*. Springer. https://doi.org/10.1007/978-981-15-1137-0_6

24. Sahu, M., Gangaramani, A., & Bharambe, A. (2021). Secured crowdfunding platform using blockchain. In V. E. Balas, V. B. Semwal, A. Khandare, & M. Patil (Eds.), *Intelligent computing and networking. Lecture notes in networks and systems* (Vol. 146). Springer. https://doi.org/10.1007/978-981-15-7421-4_3

25. Arnold, L., et al. (2019). Blockchain and initial coin offerings: Blockchain's implications for crowdfunding. In H. Treiblmaier & R. Beck (Eds.), *Business transformation through blockchain*. Palgrave Macmillan. https://doi.org/10.1007/978-3-319-98911-2_8

26. Andoni, M., Robu, V., Flynn, D., Abram, S., Geach, D., Jenkins, D., McCallum, P., & Peacock, A. (2019). Blockchain technology in the energy sector: A systematic review of challenges and opportunities. *Renewable and Sustainable Energy Reviews, 100*, 143–174. https://doi.org/10.1016/j.rser.2018.10.014

Index